QUANTITY SURVEYING

Practice and Administration

DENNIS F. TURNER

BA, FRICS, AIQS, MCIOB

Third edition

GEORGE GODWIN

LONDON AND NEW YORK

George Godwin
an imprint of:
Longman Group Limited
Longman House, Burnt Mill, Harlow
Essex CM20 2JE, England
Associated companies throughout the world

Published in the United States of America
by Longman Inc., New York

First published 1972
Second edition 1979
Third edition 1983

Reprinted 1985

British Library Cataloguing in Publication Data

Turner, Dennis F.
  Quantity surveying—3rd ed.
  1. Quantity surveying—3rd ed.
  I. Title
  624.1   TH435

  ISBN 0-7114-5756-5

Library of Congress Cataloging in Publication Data

Turner, Dennis Frederick.
  Quantity surveying.

  Bibliography: p.
  Includes index.
  1. Building—Estimates—Great Britain.   I. Title.
TH435.T87   1983   624'.1   82-12019
ISBN 0-7114-5756-5 (pbk.)

Printed at The Bath Press, Avon

Contents

PART 1—THE PRACTICE AND ITS ENVIRONMENT

PART 2—THE PRE-TENDER STAGE

CONTENTS

PART 5—TWO CASE STUDIES

CONTENTS

List of figures

Introduction

The introduction to any book presents the author with the problem of answering the question: 'Why has this book been written?' In the present case the aim was to survey the managerial and procedural aspects of quantity surveying, as distinct from the techniques employed at the working level. These aspects embrace the ways of a long established and widely operating profession, and so the task of answering the question becomes all the more formidable. It may well be said that one finds out how to practise by practising and that any secondhand preview will be of very limited value. It may also be said that when once one has practised, the preview is no longer needed.

With both these propositions I must express a large measure of sympathy. Nevertheless, students and others entering the profession do have to meet practice aspects during their days of study and early work experience. The pace of the profession is speeding up and people are 'in at the deep end' earlier than ever. It would seem, too, that the trend of the future is for more and more clients to expect advice for decision making from quantity surveyors, and not just the performance of a skilled backroom service that mysteriously produced the finished goods out of the bag, without consultation with or by anyone else. I have therefore written this book to bring together the various strands of practice that crop up in syllabuses of study and that form an intrinsic part of the background knowledge needed in tackling tests of professional competence. If others, further along the road, find the result of use in re-examining some aspect of their experience, so much the better.

While the emphasis is thus upon straight quantity surveying, I would hope that students in other disciplines will find that much of the ground covered is relevant to their own studies, which often overlap with those of that peculiar breed, quantity surveyors. Here too there may be active practitioners who will be intrigued enough to take off the lid and look in on what quantity surveyors get so complicated about.

Changes in the area of standard contract forms have led to most changes in the text for the present edition. The extensive 1980 revisions to the JCT forms have produced numerous changes, some large and many small, especially in the post-contract chapters and the case studies. In addition, the introduction of new contract forms and peripheral documents into the JCT family has had its effect. The institutionalisation of the design and build approach in particular has led to an expansion of treatment of that concept and of other ways of arranging the project team. However, I have continued to avoid discussion, except in passing, of the contract law of which practice must take account. Here the material set out in my other work *Building Contracts: a Practical Guide* covers the ground in some detail and I have referred to it directly at intervals in its fourth edition for the 'nuts and bolts' of the several contracts, while seeking to make the present text complete and intelligible in its own right.

The area other than standard documentation that has received special consideration is that of cost control and economics. Here again there is an established body of techniques in such fields as cost planning which do not fall within practice as I have delineated it. However there is an interface between these techniques and practice at which the effect of practice decisions flows through into the economic area. In view of the growing importance of client advice in this area it has been necessary to make the influence of practice explicit at several stages of the discussion, rather than implicit as in the first edition.

The contents of this edition follow the pattern of the others. After the chapters dealing with the more general considerations of practice, the sequence is an historical one, as is the way of a particular project. This allows a comparison of the various bases at each stage, which is one of the fundamentals to be grasped from a practice point of view. It also allows the traditional and still predominant system of bills of quantities to be taken as a reference point for many of the other systems, thus starting on more familiar ground. If it is desired to follow a single system through, this can be done without problems in the arrangement used. It has also seemed best to write from the standpoint of quantity surveying in the equally traditional independent setting, rather than to attempt to cover also those facets that specially concern the surveyor working within a contracting organisation. Most of his concerns are still treated by this means, however.

Some matters have received special emphasis. These include the less usual approaches of prime cost contracts and the all-in service. In the traditional field, special emphasis has at times been placed on interim valuation principles, since it is then possible to examine a number of principles that are wider than valuations themselves but that are compactly illustrated by them. Some are legal and some are practical; one that is frequently being met, unfortunately, is that of a contractor's insolvency. The aim of the case studies is to bring out a number of these issues in as near to a live setting as can be achieved between two covers. They are intended to be read where they are placed, after the other parts, but to act at examples to be worked to consolidate details. Alas, it has been necessary again to multiply up all the figures to allow for inflation!

Almost too often I have written some note of near apology to the effect that the treatment is one of broad principles only. My justification is that practice is a mass of principles in action. It is to be hoped that these are working principles and I have tried to avoid theorising on the one hand as I have tried to dodge the technicalities of quantity surveying on the other. I suspect that I have still ridden one or other of my hobby horses from time to time.

There are numerous valuable pro-formas for routine clerical operations, such as the preparation of valuation statements, which save much time and act as check lists on what is being done. While noting some of these in the bibliography, I have deliberately not employed them in the text so that the reader is introduced to general principles, rather than to a ready-made procedure. At this stage in studying, even more than before, there is a need to understand what is being done and why, and not to accept uncritically and repeat in the same way. This is how the present text should be used as well.

I continue to thank my wife for her help in many ways. It would be invidious to single out any group alone: there are students, colleagues in quantity surveying as well as architects, engineers and contractors who have suffered me patiently and whom I hope I have repaid by letting something rub off on me. The bibliography reveals a little more of my indebtedness.

DENNIS F. TURNER
May 1983

Note

References have been made throughout this book to the author's other work *Building Contracts: A Practical Guide*, for fuller discussion of contracts. These references are to the 4th edition (May 1983) of *Building Contracts* and have been abbreviated as *BCPG*.

References to the JCT Form are usually to the Standard Form of Building Contract, Private Edition with Quantities, 1980 edition, as issued by the Joint Contracts Tribunal.

References to the Standard Method of Measurement of Building Works are to the 6th edition, August 1978.

PART 1

THE PRACTICE AND ITS ENVIRONMENT

CHAPTER 1

The type and structure of the practice

Most quantity surveying practices, at least in the private firm sector, come into being in a small way. Not a few have started on a table in the bedroom and have been administered under the flickering light of midnight oil — and 'charring' has often provided the price of the oil in those early days. For the practice that survives this stage, or that bursts into being with a welter of commissions, the problem of structure has arrived—even if its type is already determined.

A particular problem arises for the smallish, growing practice over the matter of overheads. The 'man and a boy' practice carries on well enough with a part-time secretary for letters; filing hardly exists. With a staff of about half a dozen the practice needs a full-time secretary and the first human overhead has settled in. But at the same time in the case of a sole principal, and not too far ahead in the case of a partnership, another overhead arrives. This time it is the principal himself, or his plural equivalents, because at about this stage it becomes difficult for him to do more than supervise the work of the staff, administer the practice and secure its work, and in between whiles do some cost advice, editing and the drafting of preliminaries.

It is probably a happy thing for the firm, at this stage, if it is able to expand to ten or more assistants per principal. Such a number still allows personal control and knowledge of what everybody is doing, and at the same time it spreads the overhead element of the principal across a reasonable number of producers. Clearly size helps other overheads also, although there is a point beyond which it creates other problems—chiefly in the field of administration and technical data control.

Even in a practice with ten assistants the question of working groups will already have had to be faced, as discussed later in this chapter. Before considering this, however, it is useful to discuss in outline some of the main types of practice that are current. Their basic structure problems will be the same but the character of the practice will lead to differences of emphasis.

Types of practice

INDEPENDENT PRIVATE PRACTICE
This is the traditional type of firm whose historic development is

inseparable from that of professional quantity surveying practice. The practice often secures its commissions through architects with whom it has worked in the past, although the appointment of quantity surveyors on the direct initiative of the building client is a growing procedure, and one which many quantity surveyors welcome as enhancing their standing and independence. Even where this latter procedure is used, however, advantages still accrue where the same firms of architects and quantity surveyors work together over a series of projects and establish a pattern of operation.

The essence of such a practice is that it acts in a strictly impartial role in the financial aspects of the building contract. Being neither client nor designer nor contractor it does not have a vested interest to maintain and can act without fear or favour in coming to a settlement. Unless the quantity surveyor is under undue influence the system generally works as it should.

For some time there have been calls for the integration of design and economics, or of design and construction, and this has been seen as leading to the future decline of independent practices. Against the former it may be argued that the independent practice will probably work with a larger variety of designers and thus gain breadth of economic vision. Against the latter may be adduced the continuing value of an independent financial adviser to the client. It does not seem likely that any single arrangement is going to oust the others for a long time, if at all.

CONSORTIUM OF PRIVATE PRACTICES
Under this arrangement two or more firms come together to offer their combined services for a particular and usually large project or for a series of projects of a similar nature, such as may arise out of the development of an industrialised building system. The consortium will usually embrace several designers, as well as the quantity surveyor, and will aim to work more closely than under the traditional pattern—particularly where development work is involved. The client will pay a unified fee or royalty and it is this feature which is the distinguishing mark that justifies the use of the term consortium. The member firms will share out the remuneration by internal agreement, perhaps on a basis fluctuating according to their eventual contribution. The quantity surveyor in particular may find that he is not always working in the traditional position of professional impartiality; this may be a perfectly proper situation but the fact needs to be recognised.

Such a consortium is by its nature temporary. The members may be carrying on their normal practices at the same time and, in any case, will resume them when the arrangement ends. Occasionally two firms of quantity surveyors may themselves form a consortium for a special purpose, without designers or other members of different professions.

INTEGRATED PRIVATE PRACTICE

Here the consortium arrangement passes from a temporary to a permanent arrangement, by the formation of a single practice covering the design and quantity surveying functions and with its principals including representatives of each of the functions embraced. A practice of this type is able to offer a complete service and a single fee structure to the client, even though this fee structure is the aggregate of the fee structures of the various functions offered. Apart from this selling point, the system permits the closest working of the members of the team, since they are literally under one roof in most cases. It should result in better building for the client and in greater profitability and increased future business for the practice. Whether the practice achieves these aims will depend on the depth of integration at working level; the practice may still only be a series of departments, working virtually as separate entities. Success is likely to be greatest when the team for a project works not only under one roof but, as nearly as possible, in one room.

The responsibility of the partnership for the errors of the individual professions composing it will be far-reaching; this will necessitate careful drafting of the partnership agreement but should also heighten the desire for effective integration. (The old proverb about the strength of a chain being that of its weakest link is very true here.)

There are clients who wish to select particular consultants of their own and in such cases the integrated practice may supply less than a total service. The quantity surveyor, as financial consultant, is an obvious candidate for selection here. At the other end of the scale, some devotees of the integrated practice foresee the day when the practice will supply the contractor as well, by nomination from a panel according to the type of work.

The question that the quantity surveyor in particular will have to face is whether today's integrated practice affects his independence of action in relation to the contractor. He will still be 'the quantity surveyor' under the JCT Form of Contract*; to exercise this position to the full he may need at times to dissociate himself from the rest of the integrated team. His problem is similar, to some extent, to that of his public service colleague. The integrated practice needs to come to an integrated solution and then hold to it through thick and thin.

---

*Except where otherwise stated, any reference in this book to the JCT 'Form of Contract should be taken to mean the various versions of the standard form issued on the authority of the Joint Contracts Tribunal and published by the RIBA.

PUBLIC SERVICE PRACTICE

It is very unlikely that any public body will maintain a quantity surveyor's section or department without at least an architect's department, although there may well be an architect and no quantity surveyor. Where there is a quantity surveying principal he may find himself in one of several positions. In national bodies such as a statutory undertaking or board he is more likely to find himself at the head of a distinct department, on an organisational and titular equality with the chief architect if not with the same seniority of grade. In local government service he is often within the architect's or engineer's department, whatever his title may be. The differences in titles and apparent status may mean much or little so far as the type of practice is concerned; this may be such that it approximates to an independent or an integrated practice, or to some compromise in day-to-day operation for the quantity surveyor.

Like his colleague in the integrated private practice, the public service quantity surveyor does not stand in an entirely independent position. Again he is a member of a team but needs to establish freedom of action. But it goes further than this, because he is also a direct employee of the client. While this should not affect his professional integrity it may render him subject to standing orders and procedures that may limit his freedom of action. On occasions he may find himself overruled. There can be little doubt that one reason for the growth of contractors' quantity surveying organisations has been a reaction against the growth of corresponding organisations in the client's own employment. The quantity surveyor is not mentioned in the standard form of contract for civil engineering, the ICE Form, and his standing under the contract is to that degree affected whether he be an employee or in independent practice. (This form of contract is discussed in detail in the author's *Building Contracts — A Practical Guide,* Chapter 28.)

Despite these apparent difficulties, most public service quantity surveyors manage to operate in a way very similar to that of their opposite numbers in private practice and, in some respects, to enjoy a greater practical independence.

COMMERCIAL SERVICE PRACTICE

This may mean two things. One is the quantity surveying department of a building contractor and this, as mentioned in the introduction, is outside the scope of this book. The other is the quantity surveying department of a commercial organisation which is primarily engaged in some business other than building; some of the larger concerns now maintain a department to deal with their own building programme. In principle the practice is like that of a public body, in that it is in the direct employment of its client. Again, therefore, it should be noted that its independence needs to be clarified.

The structure of a practice

It has been pointed out that increasing size brings problems to any practice. This includes those that relate to the structure of the working organisation. They are discussed here in terms of the non-integrated practice, as this is the common form. The principles will still apply to the integrated practice, with some adjustment of detail.

The really small practice must of necessity operate on the basis of 'all hands to the pumps'. So far as their capabilities extend, each member of the staff will expect to do a bit of everything. There will be insufficient work for anyone to specialise, yet what is pressing at any one time may well call for all available resources. With the growth of the practice comes the need to divide up the organisation and the possibility of specialisation. These are encouraging signs and they should be exploited, deliberately but flexibly. Whatever structure is chosen, it should not become inviolable and it should not hamper the efficient working of the practice as a whole.

PARALLEL GROUP DIVISION

The simplest method of division is to split the staff into two or more groups, each with a group leader and each capable of carrying out the whole round of work that the office handles. As jobs come into the office they are passed to the group with the most spare capacity, and that group sees the job through to the end. This pattern is shown in outline in Figure 1.

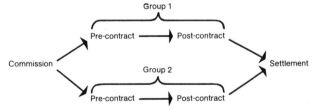

Figure 1

All that this has done is to produce two small offices; each will probably operate on the basis of the whole group tackling the larger jobs at pre-contract stage and parts of the group tackling the smaller ones, while nearly all jobs at post-contract stage will be handled by an individual member of the group. This last arrangement follows the sound principle of involving the minimum number of people in any process of negotiation.

The advantages of this structure are those of retaining small office working. There is an economy in the continuity of staff from end to end of the project, the cost planner and the final account man will both be part of the taking-off team (and will sometimes be the whole of it) and at each stage one person at least knows what has been done at the preceding stage; there is not need for familiarisation every time. The majority of staff prefer such an arrangement; they see the whole story of a job and this is more satisfying for most people than if they see only part. By the same token, they see the results of their mistakes and, it is to be hoped, profit by this. If the composition of the group remains constant over a period its members will obtain as wide an experience as the range of work permits, which should benefit both them and the office. Inevitably, however, there will not be any increase of production through specialisation. Another disadvantage usually experienced arises out of the tension between the differing rhythms of jobs at the pre-contract and post-contract stages—this sometimes finds expression in the slogan 'Get out to tender at all costs and settle up when you get the chance'.

An obvious refinement of parallel group division is to allocate each group to the work of a particular client or to work of a particular type—say housing or factories. This will give a measure of specialisation which will aid productivity and may improve client relations by continuity. According to how far this is taken, it will clearly reduce the advantages of group division to some degree. And, of course, it leaves the final account problem just as it was.

The branch office is an example of parallel group division and usually arises out of the dictates of geography as well as the growth of the practice. Occasionally the position is reversed and the branch is established to encourage the growth of the practice in a new area—but this is the realm of policy, not structure. The branch as such should work like any other group, except that it will have less easy access to the central facilities of the firm and will need to develop some of its own.

Each group and branch will be doing less than its best if it fails to help another group or branch in need. This is where the principals must keep their collective fingers on the communal pulse—and act promptly where necessary.

PRODUCTION LINE GROUP DIVISION

This structure differs from parallel group division in that it abandons the principle of making the practice a multiplication of the way in which the one-man band works. The work is analysed into stages and the practice is divided into those stages, with a group of staff to tackle each. The

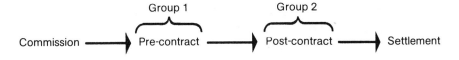

Figure 2

simplest production line will be that shown in Figure 2.

Specialisation is the keynote here and this may be increased in a large practice by dividing the pre-contract stage into sub-stages such as:

(a) Estimating and cost planning

(b) Production of tender documents

(c) Examination of tenders and negotiation

Owing to the differing rhythms previously mentioned, it is better for each group to be self-contained as far as working-up arrangements go. If this is not practicable, then working-up should preferably be a group on its own; if it becomes part of one group it is likely to be less available to the other groups. Only in the largest organisations will it be practicable to maintain distinct groups for such purposes as design and method research or cost analysis and records. Elsewhere they will be by-products of the other groups.

Production line group division in quantity surveying carries the advantages and disadvantages of specialisation as it does in any other line of business. Whatever has been said for and against parallel group division may be said here in reverse—continuity is lost in any given job and individual satisfaction reduced but, on the other hand, individual stages may be more economically performed and, in particular, the final account may have a speedier passage.

COMPLEX DIVISION

Parallel group division and production line group division are not mutually exclusive in an organisation of sufficient size. It is possible to divide a large practice into major groups dealing with different types of work and then organise each group as a production line. It is also obvious that a large office organised on the production line basis will find it necessary to divide at least the larger stages into parallel groups. Any practice with research and analysis groups will tend to keep these back-room functions apart from the rest of the office, however they may be organised.

SUPPORT GROUPS

As soon as the practice acquires a part-time typist it has its embryo support group. With professional growth there follows a series of other folk to do other tasks. There will be duplicating and photo-copying; calculating and computer-feeding; filing, the telephone, and the post; and of course those twin necessities, salaries and tea. Some of these tasks must obviously be centralised, others may be permanently or temporarily allocated to the professional groups, ana the staff with them. The principle should be that the support group is organised and positioned for the benefit of the practice and not just for the benefit of its members. He is a brave man who takes his typing pool to task: prevention is better than confrontation.

One support group needs special mention—the organisation for printing bills of quantities and other such documents. Most private firms, and many of the smaller public offices, prefer to go to an outside firm for this specialised type of work. No doubt a printing section within the office permits that close integration in time of drafting, typing and duplicating that sometimes becomes an unavoidable necessity, but most specialist firms give a very efficient service and produce a better presented document than all but the most polished internal printing sections can give. Whatever happens, the quantity surveyor should avoid the non-specialist typing bureau; one might as well pass one's drafts to the local laundry in view of the oddities of bills—and of some bureaux.

Summary

In devising or revising the structure of the practice, certain guide lines should underlie the decisions to be taken. The more important are:
  (a) to render an efficient service to clients
  (c) to operate the practice economically
  (c) to give adequate responsibility to each member of the staff, while maintaining sufficient overall control
  (d) to encourage the intelligent interest of each member of the staff in his or her immediate work and its consequences.

These guide lines all point in one direction when properly taken together and there are sound ethical reasons for following them. But even without such reasons they also provide the basis for a successful practice, measured in terms of profitability or its public service equivalent.

CHAPTER 2

The team in action and its backing

The last chapter dealt with the broad pattern of the professional practice of the quantity surveyor. This chapter is concerned with the more mundane organisational matters involved in carrying through an individual job—more particularly one of the traditional variety with bills of quantities—and the background office administration needed. Here one verges in places on quantity surveying techniques but it is an area that is broadly procedural in so far as it has an effect on the working team as a whole.

The structure and working of the team
PRE-TENDER
The old saying has it that if you want a job done properly you should do it yourself. In many ways the most efficient way of working is for a single surveyor to do everything on one job; it can also be the most satisfying, though on a large job the satisfaction may be cancelled out by boredom. So far as the cost advice stage of the pre-tender process is concerned, it may be the only safe way of avoiding a major mistake in an area where judgment is an intrinsic part of operations. But time may preclude the use of just one surveyor, whatever the size of the job, in the other activities prior to obtaining tenders.

These being the realities of life, the team becomes the obvious method for producing the bills of quantities. The team leader will be the senior staff member concerned, if such there is; otherwise a useful policy is to appoint the senior team members as team leaders on successive jobs, in rotation. If the latter course is adopted, the team leader is the obvious person to handle the post-contract stages—he has made the decisions and he should therefore live with the results.

It is important that the team leader does actually make his own detailed decisions. Most practices will have their own framework of rules; these will cover methods of dealing with architects and others, the accepted format and presentation of documents, policy regarding preliminaries, PC items, daywork, and the like, along with many other points. In addition to keeping within these rules, the team leader may be given a particular briefing for the individual job, including perhaps how to dispose his team and whether its members have any other commitments to carry through. Subject to this, the

11

leader should be in control from day to day. If he does need direction he must know the one person (and only one) to whom he should go but he should not abuse this right of appeal either by its over-exercise or by its neglect. If his responsibility is to end at a particular stage, such as before final editing or in the closing part of a negotiation, this should be known to him before he starts.

Although a team may be a necessity, its size should be restricted to the minimum. Except on the multi-million large-scale scheme it is desirable to stick to three or four takers-off, including the team leader. On this basis a general pattern for the distribution of the work (using the group method of taking-off) would be:

Primary Divisions	*Secondary Divisions*
(a) Carcass work	(i) Sub-structure
	(ii) Superstructure
(b) Finishings	(i) Room finishings
	(ii) Doors and windows
(c) Services	Each separate service
(d) External works	(i) Drains
	(ii) Services
	(iii) Surfacings and walls

These primary divisions involve the least duplication of research by the assistants concerned. If it becomes necessary to bring in further staff, then the secondary divisions may be used; they step up the research element, complicate co-ordination and editing (perhaps to the point where the team leader ceases to be a producer) and increase the risk of errors, but necessity is the mother of more children than invention. If some other method of working, such as a trade by trade analysis, is used, there are ways of grouping parts into primary sets that allow the most economical working for the job in question.

Where a contract is to consist of several buildings of even broadly similar character, it is better for the same team to work through all the buildings in turn, dealing with work with which they will become increasingly familiar, rather than allocate separate buildings to separate takers-off. This will help to obtain consistency between the resultant bills, both in methods of measurement and in descriptions. If one bill (preferably the largest) is produced ahead of the rest the points raised by it can be cleared before they get to an advanced stage in later bills.

TENDER AND NEGOTIATION

Tender examination only really allows for one person to sift the information in the short time usually available, with that person backed up by machine operators and clerical assistants. Negotiation involves more detailed work for the quantity surveyor and if the work is extensive it may be possible to divide it up, subject to adequate grounds rules and to the final agreement lying with one man on each side—who should be able to take small discrepancies into account in striking an overall balance in the closing stages.

POST-CONTRACT

Here there is a large element of negotiation and the same principle should apply as in pre-contract negotiation. An important distinction is that the quantity surveyor may be dealing with a contractor who has not been chosen for his suitability as a negotiator; the contractor may therefore be confused if he has to deal with a number of surveyors—alternatively he may try to create confusion for his own benefit. For these reasons it is far better for all price negotiations to rest with one person; after all, post-contract prices do have to be determined in accordance with the conditions of contract, while pre-contract ones are not tied to any such formula that an auditor can understand!

Measurement is best left with one assistant, for similar reasons, though extensive re-measurement of drains or builder's work is sensibly delegated to a junior—as may be those variations which are separate entities.

The working adjuncts of the team

At any stage of a project the team will need to call on other members of the staff for various assistance—if only for telephone calls. But apart from this there will also be one or two systems that the team leader will need to set up within the team; those that have a control function he will wish to operate for himself but those with more value as records, for use in any 'post-mortem' on what has gone wrong, can usually be delegated to another member of the team. A surveyor working solo on a job will still need these adjuncts but he can simplify them in part.

JOB PROGRESS CHART

This item is particularly useful in the pre-tender stage, to ensure that the job will be ready by the date planned. At tender examination it may be replaced by a check-list of work to clear, while in the post-contract stage the quantity surveyor is usually dependent on the pace of the building operations rather than himself setting the pace. In addition, post-contract work tends to be more fragmented unless there is complete re-measurement, and therefore defies efforts to reduce it to a useful chart.

The pre-tender chart is produced in its first version as an exercise in feasibility. The total work content of the job should be assessed in terms of the different types and experience of staff required. This should then be compared with the time available to carry out the work. This comparison will reveal one of several states of affairs:

(a) Division into the primary divisions referred to earlier in this chapter will be sufficient.
(b) Division into some or all of the secondary divisions is necessary.
(c) Further division is necessary and practicable.
(d) Further division is necessary but impracticable (few things are impossible!).

These states become successively less economic. Expediency will dictate whether to ask for more time or just to grin and bear it. Before going ahead, however, it must also be decided that sufficient staff can be available in the group or can be loaned to it. Other current commitments of individual staff, and the possibility of overtime, must also be assessed.

Having arrived at outline feasibility so far as the office is concerned, the team leader will now fill in the first version of the chart, a typical example of which is shown in Figure 3. This is divided into columns representing the staff available and has a vertical time scale. These are the fixed elements; the order of working through the job is adjustable. (The only advantage of a vertical time scale is that it makes it easier to write notes across the sheet and against a single date; the alternative is a bar chart with a horizontal time scale.) Into the chart is now entered the optimum division of work. In a complex chart colours will obviously be of value to distinguish different buildings or other work blocks.

The next stage is to draw up a schedule showing when design information is required to meet the programme shown by the chart. If the design programme is not known in sufficient detail the position should be checked carefully; it is vital to reconcile the design and the quantities programmes and when this has been done the team leader must be ready to revise his chart to suit the situation. If he cannot arrive at a solution, or obtain a change in the design programme, he must ask the practice to bring bigger guns to bear to secure an amendment of the design programme or the client's programme. Having come to an agreement, the team leader should confirm the dates for receipt of information; this confirmation to the designer should be in writing.

Hereafter the team leader will need to keep his chart under careful review, marking actual progress against that anticipated, taking action to rectify dangerous trends, and modifying the chart in the light of developments. Charts can become an obsession, and a great deal can be carried in the head instead of being elaborately charted, but a simple chart is objective and easy to prepare and is then available to all concerned.

Project: Doll's Hospital

W/E	TOM	DICK	HARRY
9 May	} Substructure Piling Sub-bill	} Finishings (except Lab wing)	//// Not available ////
16 May	to go out by. }		////
23 May	} Frame Visit Site	} Doors and Screens	} Roads etc.
30 May	} ? Walls Key	} Windows & Roof light	} Drains & mains (use Jim here as well?)
6 June	} Prelim queries Upper floors	} Lab wing finishings Margin	} Services
13 June	} (inc sub-bills) Roofs	//// Holidays ////	}
20 June	} Prelims	////	//// Valuations //// } Services complete
27 June	} Preambles Tidying up	} supervise final billing	} Fittings
4 July	} Editing	} Jo print	} Help billing

(Allows three days in hand)

Figure 3: Taking-off programme

Projects vary so much in value and character that experience is the only real guide to the surveying time content of any individual job, but although no book formula or schedule is of real use a few general points may be made. Time required is not directly related to value; it will fall off proportionately as value increases and even more so with repetition. The character of a job also affects the issue, quality and complexity here having the reverse effects on the time and value relationship. Where time always seems to have been under-estimated is at the end of a job; there must always be some slack in the programme to allow for over-optimism; or the short-comings of one's own staff or that of the designer. But in addition to this, the editing stage should be assessed generously to allow for all the unforeseeen matters that accumulate.

Adequate allowance should also be made for tendering time, examination and placing, and the time needed by the contractor before starting on site. These matters are discussed later but it should be said at once that it is a common fault for the programme to cramp them; overall they require at least two full months even on a modest scheme.

REGISTERS OF INFORMATION RECEIVED
These will be made at all stages of a project and will cover drawings, schedules and written information. In pre-tender work they serve to show whether information has arrived on time and will also act as an index to what was used in the preparation of the quantities. Later they serve to index information which may come over a lengthy period but may eventually be used all together in dealing with variations. If the information itself is not fully dated the index can be used in the event of claims, to verify at least the approximate dates on which the contractor received the information.

QUERY LISTS
These lists are very useful in the pre-tender stage and usually consist of two columns, one for questions and one for the answers. One cumulative list can be kept for the whole project but it is better for each taker-off to have his own, starting a fresh sheet for each section of the work. The rest of the team should see each member's lists regularly, in case their own work is affected by the answers. Copies of the lists should be sent to the designer at whatever intervals are necessary to ensure that later information does not run counter to the answers. While the bills are out to tender it is a good plan to meet the designer and make sure tactfully that all query points have been embodied in the contract and initial issue drawings.

In the tender and post-contract stages letters are usually better means of confirmation than query lists; they are, of course, also useful at pre-tender stage.

The supporting organisation

What has been said so far in this chapter has related to the passage of a single project through its various stages in the practice. The description is therefore appropriate either to the whole of a small office or to a section within the larger office. In either case there will be parts of the organisation that are there not to contribute to a single job, but to keep the practice ticking over. How these parts are organised will depend very much on their size and is not for present discussion, any more than is the detail of the equipment which they tend to attract to themselves at an alarming rate. There are some general divisions into which the systems will fall, however.

THE GENERAL FILING SYSTEM

Few things cause so much frustration or lead to so many expert opinions on how it should be done as the office files. If there is a pan-office system, it may become the aim of whoever institutes it to absorb into it every piece of paper that is not actually being written on at any given moment. The reaction of individuals, for a multitude of reasons, is to keep everything 'where it can be found'. There is distrust of the central system or of every other user of it, a desire to arrange things according to other criteria for easy retrieval, or simply a disinclination to have to walk about the office. In any medium-sized practice and above, some compromise must usually be found.

Where individual surveyors or teams are engaged on a particular project for any length of time it is obviously necessary for them to keep the technical information that they receive or produce close at hand, and to have control over its classification. This information will include drawings, estimates and dimensions, to give three common examples, and how these are kept may vary according to the stage of the project. But even though considerable latitude should be accorded the member running the job, there should be an established office framework that is observed, so that it is reasonably possible for someone outside the immediate team to find at least the more vital documents in an emergency when the team happens to be absent. This is especially true of financial statements and their supporting calculations, since it is here that an architect or client is most likely to want quick explanations of what he has been sent. No system can make up for the living persons who operate and understand it, but it can help.

Correspondence is a peculiar bone of contention on the score of whether it should all be kept centrally or with the other papers of the job to which it relates. Certainly much of it is intimately connected with the running of the job from day to day. To separate letters on a job to either the central files or the job files according to subject matter may be to court fresh confusion, even though it must be done in finer detail at the level of either the central system or the job file system. Where there is any reasonable doubt it is best to keep a copy in each place.

When a project is finalised in its various aspects, the papers should be sorted and those of importance put into storage files. Two main considerations will affect what is to be kept and what discarded—apart from the creditable aim of not cluttering up valuable space. One is the matter of what is likely to be of value in the future working of the practice; this will include cost information extracted at tender or final account stage, as well as extracts from, or sometimes complete copies of, the bills of quantities or other documents and the supporting drawings, quotations or the like. Some of this data is likely to be needed as ready reference information and is best concentrated in the technical filing system so that it may be referred to by anyone in need. If then they find that they need to go behind the tabloid data, as is always wise if in doubt, they will refer to the main papers grouped under the job reference in the old job files.

The other consideration affecting details from completed jobs is that of the surveyor's responsibility to the parties to the building contract, should they have any query of substance. This is immediately a responsibility to the client with whom the surveyor has his own contract, but may be related to any point that the contractor is entitled to raise to the client. Under the provisions of the Limitation Act, this means a matter of six or twelve years from the ending of work on site for physical matters and these may carry financial items in their train. For whichever reason papers are kept, they should be reviewed from time to time to see whether they can be pruned or eliminated.

The general administration of the business side of the practice will require a centralised section to the system, apart from the correspondence files alluded to above. Much of this will be in support of those items that are looked at below as the accounts that the practice needs. Even when these files are not confidential to particular members they should be carefully regulated, since there is nothing easier than to lose a central system by wastage.

THE TECHNICAL FILING SYSTEM

If the last remark is true of the general central system, it is the first and last consideration in setting up a central technical system. Such a system should be as simple as its contents permit, it should aim to be accessible and to control by iron chains whatever it lets out of its own self into some other system. At the same time, it must be a lending and not just a reference library, so far as most of its contents go, if it is to be of practical use. This problem is at its height for the medium-sized office that cannot afford a full-time librarian but which is large enough to make it laborious to track down a missing item.

Three basic groups of documents will constitute the main stay of the system. First, there will need for cost analysis information. Some of this is obtained from outside the practice, from such sources as the Building Cost Information Services provided by the RICS (there being one each for new work and for maintenance) and technical journals providing similar information. This gives breadth of view, but is not so easy to interpret as the data produced from jobs carried out by the practice. This may be produced and recorded as a regular procedure for every job and then kept centrally. The alternative is to extract information from a past job as and when it is required. This means that the data extracted is fresh in the mind of the person securing it and perhaps more intelligible. Against this is the element of delay and the possibility that the person who handled the contract may not be available to interpret a mass of papers. But even with this latter procedure it is desirable to keep a synopsis of each contract in sufficient detail to permit a quick examination of whether it is likely to be of assistance if unearthed.

The second group needed is that of standard office documents. A small office in its early days will not have, or need, this group but should have an eye to forming it progressively. Much of the wording in documents like bills of quantities is the same from one job to the next, or at least should be based on common principles. In the case of preliminaries and preambles this justifies standard drafts that may be edited for successive projects. There should be a store of these, to be used in preference to taking an old job and perpetuating its mistakes or antique points, with a master copy of each document on which revisions are noted pending a completely new edition being issued. For some parts of the more common documents used, the same principle may be applicable, although often a standard library for reference may be better for individual descriptions and other wording that is put together in varied combination. There are on the market several such documents for bills of quantities and specifications that may be taken over bodily by practices or adapted to suit their individual tastes and needs. Some of these libraries are directly designed for computer coding, others can be adapted to it — none absolutely require it.

The issues of copies to each member of staff needing them is more satisfactory than a centralised arrangement for documents in the standard library category. Pro-formas for common items such as valuation statements may be kept centrally for issue as required. Some practices will keep available a copy of every bill of quantities that is prepared, so that jobs with special features may be consulted for later work. This is useful for future work on the same site and so on, but may be as well served by the past job file system. Much will depend on the numbers of bills involved and the pattern of work in the office—and the space available. Any really key features should find their way into the standard documents system or into files devoted to special features, where they can be classified.

The last group is that containing trade periodicals and other publications. Some of these are of primary interest to staff when they are considering a move and are therefore read from the back forwards; the principals have a certain risk in displaying them at all! Nevertheless, a relevant selection is important, perhaps with the copies initially in circulation. Aims to extract and file information from these periodicals sometimes fall by the wayside, but are worth pursuing for the more relevant contents. An office scout who searches for topical items that should be obtained is also a valid appointment. A number of copies of the regulations and rates and the like for the various sections of the industry should be kept centrally, even though many assistants will have their own. Standard price lists and classifications for materials, hire schedules, daywork definitions, weight lists and many more sets of data can well be treated like this and, in particular, should always be kept up to date.

Within the last group come a mass of publications that offices accumulate sometimes without the asking, in the form of catalogues of varying usefulness from manufacturers and others. Some of these are related to a particular contract, but the great bulk will be kept for their general applicability, if at all. In a small practice a quite simple personally-devised method may be the answer, but even here it is not long before a logically based all-purpose system becomes a necessity. Here a standard classification is very useful, and the CI/SfB version has gained wide acceptance. It has the advantage that it is used by many producers in labelling their catalogues, which means that routine filing can be delegated to someone non-technical if required. With all this information, there is again the problem of keeping it up-to-date as well as comprehensive. So far as prices are concerned, it is necessary to obtain quotations each time if any real reliability is needed. There are organisations that will take over the basic responsibility for a technical filing system; this is a reasonable arrangement where a larger practice cannot find time, provided the organisation can really cover the need of the practice in terms of product variety.

All of this may suggest a formidable mass of detail and this is true. Three points are worth remembering. Firstly, obviously useless information should not be let in (unfortunately what is thrown out today is often needed tomorrow). Secondly, the enemy should be divided — that is, there should be sufficient logical parts to the system and knowledge of what they are and what is in them. But thirdly, links should be established between the parts so that they can support each other and thus become allies of the practice. It is here that the various published systems can be useful as a frame and indeed some are being developed so that they link together. No doubt developments in retrieval and communication systems will take things much further in this direction.

THE ACCOUNTING SYSTEM

The finances of a practice require a number of accounts to be kept, even if solely to satisfy the tax man or the accountant. Beyond this, they also give a measure of control by showing what has happened, so that the principals may learn by experience. If they are able to do so as soon as the underlying events have occurred, so much the better. The mechanics of accounting are not the immediate concern: here discussion aims to highlight the objectives to be achieved.

For each contract undertaken it will be necessary to keep both sides of the balance if efficiency is to be watched. On the one side this means all items chargeable to the client, whether by way of fees or expenses. On the other side must be allocated the salaries and direct expenses incurred, which expenses may not all be those chargeable as such to the client. To arrive at these details there must be an office allocation system, so that each member of the staff may apportion his own effort and expenses. This means that what is recorded in the other accounts described below must be transferred under other heads here. The profitability of each job and of each assistant or principal may thus be kept under review, although it must be remembered that not every job can pay and that the individual surveyor may be faced with quite unprofitable activity. Fee scales are inevitably somewhat blunt instruments.

It follows that accounts covering the staff will be needed. These will cover all direct payments to them, such as salaries, overtime, bonuses and expenses, and for these items the factor governing their presentation may well be the requirements of the taxation authorities. If the practice is so bold as to operate a genuine profit-sharing scheme, as distinct from a productivity-bonus scheme, then additional records will be needed here and should be available to the staff. Such systems are rare, since they should be loss-sharing schemes as well!

There are many expenses of practice that are quite general in their incidence. These include large overhead items like rent and other expenses relative to the premises, and the corresponding costs of major equipment, such as printing machines, where hire or depreciation are involved as well as operating costs. There are also a number of items that are built up of a great number of individually small items that in theory can usually be allocated to individual jobs. These are such matters as telephone calls and rental, postage and petty cash. It is, however, out of all proportion to attempt the labour of allocation and so the total cost is spread over those items which can easily be allocated. With these impersonal expenses will go too the salaries of the supporting office staff involved. Where cars are provided by the practice for the use of principals or assistants a separate set of accounts is usually kept, since these cars are a sizeable expense in themselves and their cost allocation will vary. Some mileage will be chargeable to clients as expenses, some to the practice in general, and some may be allowed to staff for private purposes.

Somewhere in the inner sanctum there will reside the principals' accounts. Each will have a set corresponding to those for each member of the staff, showing any salary paid before the question of profit comes to be considered. Expenses directly incurred will be shown also. Beyond this, much depends on the terms of the partnership, if such the practice is. For this purpose an associate is an assistant, since he is not committed as a partner is. Partners proper usually contribute capital as well as contacts with clients, otherwise termed goodwill, and their own efforts from day to day. Each of these elements calls for reward and the original partnership agreement, or any modification to it to take in new partners, will spell out the details of how the proceeds are to be shared out, after making allowance for retaining a prudent margin in the business. In times of little fresh work, this margin may be eroded quite rapidly and therefore should not be made too low. In the end it is to be hoped that practices will succeed and that a due reward will come to their principals as a result. If not, there would seem to be little point in the present volume, amongst other things.

CHAPTER 3

The quantity surveyor's relationship with the client

The building client occupies a peculiarly important place in the pattern of things. If he were to cease to exist, then so would the whole construction industry. He commissions the building or other project and may well be its owner as well. He provides the financial consideration for the building process and also the site upon which it takes place and the legality for it occurring, in terms of planning permission and the like. He is one of the only two parties to the building contract, but at the same time is the employer of the professional team that will usually stand outside the contractor's organisation. Indeed, some members of that team may be directly on his pay roll. All round, he is someone of considerable moment on the scene: he pays the piper and he is the customer. Yet he too is a man under authority for he has to take account of the various statutory bodies that exercise control over so many aspects of life in building work today. Again, while he employs the team they in their turn are subject also to the authority of their own professional codes of conduct and at some point may need to stand against him if he brings undue pressure to bear in the wrong way.

In the preceding paragraph, the client has been referred to as 'he', as he will be throughout the rest of this book while his strict title under the JCT form, 'employer', will be used where the setting suggests it. But the client may be something other than 'he', even allowing for the normal convention that includes the female with the male for these purposes. The client may often be a board or committee or some similar corporate entity, perhaps commercial, perhaps public or perhaps a private or charitable concern. Where the client is of sufficient substance, there may be a technical executive acting for the whole body for most day-to-day decisions and this is usually a happy arrangement in principle. He may even enter into the building operations by acting as project manager. At the other extreme some clients seem in practice to disintegrate into a series of equally matched but corporately ineffective departments, between whom the outside consultant roams plaintively and the astute contractor drives the metaphorical horse and cart.

According to the nature of the client, the surveyor and the other team members may find their dealings affected in various ways. Thus a local authority may be acting as an agent for some other body up the line in the hierarchy or be controlled over the exercise of financial policy. The whole investment policy of national corporations may be subject to crosswinds over which they have comparatively little control. Such things may add distinctive flavours to a surveyor's work and cannot be considered in a general treatment. There are some matters that will, on the other hand, be common to any case and these are the ones which will be considered in the present chapter.

Appointment of the quantity surveyor

It is the recognised state of affairs that the quantity surveyor should follow the professional pattern of waiting for work to come to him, and not go out and seek it. His appointment may therefore be brought about as the result of a direct approach from a client or, as is frequently the case, it may be due to the recommendation of an architect with whom he has worked and who seeks to preserve an established system of working. There are obvious advantages to the surveyor also in working with someone whose habits are familiar to him, so that for any fresh project taken on it will be easy to set up the team needed. Many technical points will also be cleared in advance by knowing the other person's practices. Against this, there is the greater independence and standing that the surveyor has when he is someone whom the client selects for himself. There is something of an increase in the practice of appointing the quantity surveyor direct and many would welcome it.

It follows from the professional code that the surveyor is not to advertise his services. He may do so in terms of a direct notice only when he offers his services to his own profession, and then only in the discretion of a box number in a trade journal. He may not approach any client directly seeking work, except that where he acts regularly for a client he may enquire of that client whether there is further work in prospect, so that he may plan his programme. This principle extends to the panels of surveying firms that are maintained by many public bodies. Here the surveyor may not ask to be put on a particular panel entirely on his own initiative; he should either wait until a body advertises its desire to receive applications direct, or he should put his name on the list at his own professional body's headquarters so that interested authorities may select him from the list if they wish.

In public places, the most that may be done is to display, on a site that is already being dealt with, the recognised pattern board giving the surveyor's name and practice address. Notice in the professional press may go so far as to give restrained details of changes of address or of changes in the partnership. And here it stops: embossed diaries and other materials definitely transgress what is reasonable. The aspiring practitioner may wonder how it is ever possible to start at all — nevertheless some do and some of those become established.

When the surveyor does find himself with the offer of a commission, he may find that it is purely oral. It is not a matter of legal necessity to establish a written agreement; one that can be proved to exist orally is enforceable. Such proof may be by implication from performance: if a client instructs his architect to obtain tenders for a project of such a size that quantities are the normal requisite, then he will be held to have required the appointment of a surveyor to prepare them. This will be so even where the architect alone directly engages the surveyor. But such a situation is far from ideal and it is only marginally improved by such devices as implicating the client by correspondence that shows that something is being done. There may be times when the politics of the case may suggest some such line of action but it is far more straightforward, and in the interests of all parties, to have a properly constituted written agreement.

There are two ways of entering into this. One is by a specially written letter setting out the details of services and remuneration, this being complemented by a confirmation from the client that what is put forward is acceptable and comprehensive. This does give a somewhat personal touch that may still be valued by some today. On the other hand it does make for much detailing with the increasing complexity of the services that may be offered and of the fees to meet them. The alternative then is to use the special form of appointment that is available from the RICS, and which acts as a check list and gives the details in a properly developed manner. This can still be accompanied by a more individual letter, to keep the personal atmosphere. A copy of the appropriate scale of fees may well be sent with the statement of fees so that the client can see that they have a respectable background. It is this sort of minor expenditure that leads to a businesslike impression without fuss.

The scales may now be considered.

Professional charges

In common with many other professions, quantity surveyors charge for their services in accordance with scales of fees published by their main bodies. Members of those bodies are expected in general to abide by the scales, although they are not mandatory. Quoting any variation on these fees in competition with other members to secure work is not permitted, nor is quoting a damagingly low level of fees. Equally however a surveyor may quote a higher fee for a project involving some particularly intricate working on his part or, say, special geographical problems. Scotland has scales quite distinct from those applying in the rest of the United Kingdom.

While these scales exist for surveyors to offer to most clients, various official bodies have negotiated special scales to suit their work. There are thus local authority scales and central government scales, of several varieties in each case and published as such. In addition some other bodies have scales that have been agreed but that are not so generally published. As might be expected, these various bodies have negotiated somewhat lower scales by virtue of their own size and bargaining power, although the phenomenon may be explained otherwise. At the time of writing the whole question of scales of fees lies under the shadow of consideration by the Monopolies Commission.

The most intricate of the general scales is Scale No. 37, intended for the whole range of building work. For firm quantities jobs it provides for the cost-planning and other pre-contract stages, and the post-contract stages to be charged for separately and thus for the surveyor to be engaged to deal with only parts of the history of a contract. For the bulk of the services a percentage basis is provided and in the post-contract part there are alternatives, one more simple than the other but perhaps not so precise. As not all building work is by any means comparable in terms of capital cost or the amount of effort or skill that it requires from the quantity surveyor, the scale allows for this by two variable factors. One is a sliding percentage scale, dropping as the value rises, while the other is a broad division of work into categories according to complexity; these two factors are both applicable to any one job. There are other adjustment related to services and alterations work, while defined expenses are chargeable as extras. All of these features lead to a fairly lengthy fee account, even though it has the merit of reflecting what the surveyor has done in a number of individual pockets. There are separate parts of the scale that deal with approximate quantities schemes and prime cost work.

These last mentioned parts can hardly be said to contribute to the intricacy of the account that any one client will receive. However, because there are those clients who prefer an all-in presentation, a scale giving a single percentage applicable to a particular job has been published rather more recently. This is Scale No. 36 and the one percentage is to be used whether the job is to be dealt with on the basis of quantities or some other contract arrangement of the types described in Chapter 7. It covers all stages of the project and can therefore apply only where the surveyor is to deal with all of them. It is also limited to contracts exceeding £200,000 in value. It still uses a sliding scale according to size, and a division into categories according to complexity, but the overall result is a considerable simplification. The total sum chargeable is likely to be higher, but the all-in basis does leave the surveyor much more in a position of having to take 'the rough with the smooth'.

The other general scale is Scale No. 38 for civil engineering work and this is altogether simpler, while following the broad pattern of Scale No. 37.

Even this sketchy account, with all its omissions, emphasises the point made above—that it is important for the surveyor and his client to be clear as to the services desired and offered, and as to the remuneration to be given for them. The surveyor should expect normally to take the initiative over this, and certainly he should state his charges clearly to any client who is not used to have building work carried out or with whom he has not dealt. An initial letter giving the gist of things may be helpful before the client has decided on the services he requires.

Professional negligence

Along with the question of fees goes the possibility of professional negligence if there is a particular type of failure in the earning of that income. The present section is not a legal exposition, but a pointer to some of the areas the subject covers.

It may be noted that there is no special category of negligence known in law as 'professional'. Negligence in general arises when there is a duty to take care and there is a breach of that duty, with the result that someone to whom that duty is owed suffers a loss that is actionable. Negligence may arise in contract or in tort.

So far as the quantity surveyor is concerned, this duty is likely in the main to be one under his contract of service to the client. This contract is of course quite distinct from the contract between the client and the contractor, although the quantity surveyor has certain duties under that contract as one of the named persons (at least under the JCT form) for its proper operation. Failure in those duties may lead to an action for negligence, but this will be founded in the end on the contract between surveyor and client, which gives a contractual duty towards the client alone of the parties to the building contract. If the surveyor fails under the building contract in such a way that the contractor suffers some loss, the action of the contractor would be against the client as the only other party to the contract. The client would then be left to seek his redress against the surveyor under their mutual contract.

It was held until recently that the architect and the quantity surveyor, when acting in their roles under a building contract, were often exercising a quasi-arbitral function and in such instances were not liable for negligence, since this is not a matter for which an arbitrator is liable. The cases in question were those in which the professional adviser was acting to decide matters between the parties, and most particularly in issuing a certificate or in preparing data for the issue of a certificate. This view was founded in the decision in *Chambers v Goldthorpe,* but was overturned in 1974 by that in *Sutcliffe v Thackrah and others* so far as architects are concerned, and in *Tyrer v District Auditor for Monmouthshire* in respect of quantity surveyors (see *BCPG* Table of Cases). In these cases the professional person was liable to the client, but it appears likely that in future he could be found liable equally to the contractor. It would appear therefore that the law does not recognise any quasi-standing and that only a person acting as an arbitrator under a properly constituted arbitration appointment can rely on immunity.

It is a most remote consideration that the surveyor could suffer under an action for negligence in tort arising out of his actions under the building contract, whether this were an action by the contractor (whose remedy has been mentioned) or by some more distant person. It is however quite conceivable that he might find himself under attack when carrying out other activities. Thus in giving advice to a client he might indirectly provide to some third party information on which that third party might rely, and to his detriment if it were unsound. This principle was established, as it happened in the field of banking, by the case of *Hedley Byrne and Co Ltd v Heller and Partners* in 1963. Here a reference was given 'without liability' and this rider saved the bank giving it from suffering damages under an action in tort. The editor's headnote in the *All England Law Report* summarized the position as follows:

> "If, in the ordinary course of business or professional affairs, a person seeks information or advice from another, who is not under contractual or fiduciary obligation to give the information or advice, in circumstances in which a reasonable man so asked would know that he was being trusted, or that his skill or judgment was being relied on, and the person asked chooses to give the information or advice without clearly so qualifying his answer as to show that he does not accept responsibility, then the person replying accepts a legal duty to exercise such care as the circumstances require in making his reply; and for a failure to exercise that care an action for negligence will lie if damage results."

Liability towards the client is potentially present for the principal in practice who offers himself directly to the client. If the practice is a partnership, then the partners share the responsibility for one another's negligence whenever it arises out of a commission taken on by the practice as such. The status of a consultant perhaps needs legal definition, now that that particular species is in the ascendant numerically. Where there is a multi-professional practice, the partners are still jointly liable, even where they may not be fully competent to judge on the matter of default. In general, other members of the profession find themselves in the position of salaried employees, whether they are in a private firm or in a department of some public body or commercial concern. Associates of practices are simply employees at law. Even a self-employed surveyor working for other members of the profession is most likely to be in the same position. A practice delegating work to others will remain liable for that work, even if it is specialised, unless its contract with the client excludes the liability expressly.

What then is professional negligence for the quantity surveyor in particular? Like all professional men, he holds himself out to the public to possess the standard of knowledge and skill which is normal to his profession. He does not claim that he is the most eminent practitioner in the line—indeed if he does he may find himself facing action from his professional body for improper conduct. But he is assumed to have attained this normal standard and also to exercise reasonable care in its employment. Inexperience in the profession or in some specialised aspect with which the surveyor does not usually deal is no defence. Thus in *Brutton v Alfred Savill, Curtis and Henson* (1971) there was held to be liability when a trainee fell for a confidence trick and in *Freeman v Hall, Pain and Foster* (1976) when a surveyor dealing in valuation was found deficient in structural knowledge.

This does not mean that a surveyor is never permitted to make a mistake. The test is 'what would a reasonable person have done in this situation?' Human error may occur, but was it of such magnitude that another would have noticed the mistake before it passed beyond recall? The advice given was not the very best as it turned out, but was it a reasonable assessment of the factors as seen beforehand? If these questions can be satisfactorily answered, then there is not negligence. In the end the question becomes one for the courts and it is difficult to express precise criteria. Frequently the view is taken in such areas of specialisation that a professional man must be judged by the standards of the profession, rather than by some abstract standard set by legal principle. There would appear to be something of a blend between the intrinsic skill and care on the one hand and the magnitude of the result of the failure on the other.

Some of the more likely grounds for an action for negligence would seem to be either directly financial or those that lead a client into courses of action that result in financial loss. These would include carelessness in estimating or other processes of financial control, either in the pre-contract or the post-contract stage. Allied to these is the case of gross error in the quantities not discovered until work is advanced, or skimping of the checking of a tender, or some related aspect of the tendering firm. Any of these could lead the client into expenditure that he could not sustain or that was unacceptable. A further ground might be that of failing to recommend paying adequate or prompt interim sums to the contractor, thus leading to action against the client. The other side of this coin is to recommend payments which are more than adequate and too advanced. These will seldom lead to complaint unless the contractor stops work, particularly due to insolvency, but then their effect is most noticeable.

From all this it may be concluded that the surveyor, when dealing with quantities at least, is in a less vulnerable position than the architect, as the latter faces the array of responsibilities that are his over survey, planning, design, recommending firms, supervision and expenditure. This is fair comment but the quantity surveyor, like other professional gentry, may still find himself foregoing part or the whole of his fees and possibly subjected to an action for damages as well. The level of such damages will be that which covers the loss that could reasonably be expected to flow from following the advice given, or otherwise suffering the consequences of, the surveyor's act or omission.

The question remains what steps can a surveyor take, either before or after the event, to cover himself against a charge of negligence. Recent legal decisions mean that he can no longer claim the protection of quasi-arbitration; nor can he escape liability for advice used by those other than the direct recipients. It is also no defence to say that advice was given gratuitously — as the *Hedley Byrne* case illustrates, it is enough to know that you are being taken seriously for liability to arise. There is the option of giving advice in suitable cases to a client with a specific disclaimer of responsibility to other persons gaining access to such advice. This may be quite justifiable, since advice given to one person may be misleading to another unless set in context or qualified. But often it can undermine the confidence of a client in his adviser, unless it can be made crystal clear why it must be given. In any case the option does not affect many instances, particularly the bulk of those with which the quantity surveyor is faced. A better safeguard when dealing with unusual or ambiguous commissions is to ensure that the brief is clarified, that the terms of engagement are specific as to scope and that the advice is qualified as to its limitations. Any delegation of work involving delegation of responsibility needs care.

What a quantity surveyor cannot have is the protection of limited liability. Equally he cannot rely on immunity from an action for negligence such as a barrister enjoys. Yet his professional activities may on occasion put him at risk if he is to act with the same positive strength as the barrister. Here the remedy lies in professional indemnity insurance, not as a cover for rash or irresponsible action but to allow reasonably courageous action in suitable cases. Although premiums are quite expensive, such insurance is something which no practice can afford to neglect, out of concern for itself (both partners and staff) and also for its clients. The precaution for any surveyor or practice is to possess the skill and exercise the care in the first place.

CHAPTER 4

Relationships within the design and construction team

At the beginning of the preceding chapter it was stated that the building client was in a direct contractual relationship not only with the contractor but also with the various design consultants who might be involved. In addition, the contractor will be in a contractual relationship with a number of firms, either nominated to him or of his own choosing. These firms will not however be in any direct relationship with the client, any more than the designers have any direct relation with the contractor. So far as contracts are concerned, the pattern is something like that shown in the upper part of Figure 4. It is the client and the contractor who, through the building contract, provide the connection for the rest. If the architect engages consultants entirely at his own discretion, and therefore privately, then there may be a slight re-arrangement of the pattern, but it will also be to the left of the client. Only where the contractor or a sub-contractor provides some design service is there likely to be any disturbance of the left to right division. What this effect would be is not easy to say in round terms; it would often be untidy—which is one of the reasons for the employer/sub-contractor agreements and the supplier warranties that are used. (See *BCPG*, Chapters 17 and 19.)

There is another set of relations, other than the contractual, that exist on a project. This set is for procedural purposes and it is shown in the lower part of Figure 4. There is still a left to right division but the architect has assumed the link position held by the client before, while the quantity surveyor has taken up a position to one side of the rest. This is the situation embodied in the JCT Form of contract, for example, and results from the division of design and construction that largely prevails throughout the industry. So far as procedures go, it aims to make the architect the only person responsible for instructions and able to give them directly to the contractor. Even the client may not do this. The quantity surveyor alone is free to have direct dealings with the contractor and does not have to deal solely through the architect. This is because his role is to record and evaluate and not to design and instruct; so far as he influences these latter activities it will be by advice to the designers themselves.

In this chapter some of the key points, for present purposes, of these two sets of relations are mentioned. In addition, one or two other persons are mentioned, who have been excluded from the diagrams for simplicity and

(i) Contractual Relations

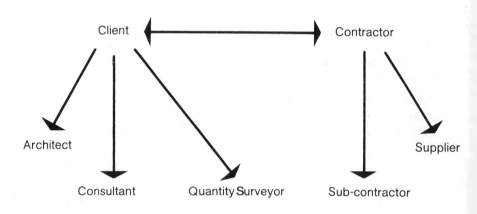

(ii) Procedural Relations

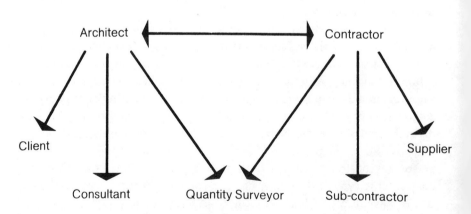

Figure 4: Traditional design and construction relationships, contractual and procedural.

because they do not affect the main philosophy of them. Since the surveyor's dealings have a particular significance so far as the contractor is concerned, something is added about his particular workings. Whatever is immediately relevant about the quantity surveyor's relationships is mentioned under the headings concerned. Except in matters of cost and contract advice, he tends to follow his traditional function of coming along after the Lord Mayor's Show—and this is the burden of much of this volume.

The employer's advisers
THE ARCHITECT
The title 'architect' is the one used in the JCT form, although other standard forms have a corresponding expression. (The person so named in the JCT form need not be a registered architect but any person who, not being on the statutory register of architects, allows the term to be applied to him more widely in any other contract document or elsewhere may find himself in trouble with the Architect's Registration Council.) Usually he will operate as the primary member of the design team in two ways: he will be appointed first out of the members, and may often advise on the selection of the rest, and he will act as the integrator and leader of the efforts of the rest. This is traditional and still the most common arrangement, although it is more and more frequently being varied in these days. It is particularly affected if the client has his own project manager or other technical executive, who does some of the supervisory work which was formerly the architect's prerogative. This is discussed further in this chapter.

The architect's more common duties are to interpret the client's brief or help him to decide it, to prepare the preliminary and detailed designs, and to supervise the construction of the works and see that they can be accepted as satisfactory by the client. Alongside these, he is responsible for the overall economy and suitability of the scheme, even where he has delegated certain aspects to others. He may also perform a number of incidental duties, such as a survey of existing premises, obtaining planning permission, and assorted legal matters that do not fall to the client's solicitor. In all these duties he is acting under his contract of engagement with the client as a professional man; in some of them his function may also be that of agent. In some of them he must exercise the powers and responsibilities that he holds under the building contract in which he is named, but to which he is not party. This sometimes puts him in the position of quasi-arbitrator between his client and the contractor where there is a matter in dispute, but not otherwise.

Towards the contractor, his standing is that he alone may give instructions, as work proceeds, about the works which he has designed as architect in the first place, although this power does not extend to directing the contractor over his site organisation and methods. All consultants should channel their instructions through him, both so that he may co-ordinate the design and to avoid administrative confusion. It will usually be the case that the architect briefs the consultants on their work for similar reasons, unless they are performing some specialised function that can best be discussed directly with the client.

The quantity surveyor, too, will be briefed by the architect in most cases, unless he has had any prior dealings with the client on the economics of the scheme. He will be adviser to the architect on the finances of the project as and when he is asked. To the degree that he is asked for financial advice he will become responsible for it, but the overall responsibility will be with the architect. There are also some matters of contract settlement that only come the way of the surveyor as and when they are delegated to him by the architect.

THE CONSULTANTS

It may be that the architect undertakes every part of the design himself, but this is infrequent where work is on any scale. Even with the smaller job, there will often be elements for which a specialist firm will give a 'scheme and a price' and thus perform some design. But on the larger project it is common to find one or more consultants employed, that is to say specialist designers paid by the client but under the general direction of the architect. Their services may embrace a small part of the work, such as a decorative feature or a limited acoustic treatment, but they are often responsible for the main structural design and the major services installations. There is usually some diminution of the architect's own fee on account of their employment, but the result of employing them is an overall increase in the fees payable, reflecting the greater complexity of the work being designed. Where the service is supplied by a firm that is to become a sub-contractor the capital cost of the works includes the design element on this part and the architect receives a fee on it; this is permitted, for restricted items, by his fee scale.

The main point about an independent consultant is that his procedural relationship with the contractor, or with the sub-contractor whose work he is designing, should be through the architect, who in turn will deal only with the contractor. It is to be noted that the standard contract forms recognise only one architect or his equivalent; any consultant is absorbed contractually, so far as the contractor is concerned, into the architect. While the sub-contractors, if nominated, have a separate existence they have no direct relation with the architect and so again it is the contractor who should receive instructions to pass on to them. The only cases in which it may be reasonable to relax slightly the letter of procedure is if some matter presents no point of overlap with the contractor's own work, such as the size of a steel bar within a concrete column. Even here there should be proper confirmation of what has been done direct to save time. It must always be asked whether in fact there is going to be saving or just the risk of confusion.

THE CLERK OF WORKS OR THE RESIDENT ENGINEER

The terminology of the JCT form speaks of a clerk of works who is the client's man, engaged to inspect the works for quality but not to give any directions, either by way of variations or, it must be held, by way of condemnation of anything defective. He is however to operate under the control of the architect and acts as an extra pair of eyes for him. Whether he is given any delegated authority is a matter to be decided on the particular contract, but if so it should be carefully defined to avoid the danger of an instruction subsequently turning out to be invalid. Otherwise he can give no instructions and the surveyor may be in trouble if he accepts what the clerk of works has issued as sufficient authority. The position is no different if the clerk of works is on the staff of the architect, so long as that is his title. A site architect is an extension of the architect and here the position is different. If there are specialist clerks of works acting for consultants there will be special need to clarify which category they fall into, and to ensure that their communications are correctly channelled.

The government form GC/Works/1 recognises only the SO (Superintending Officer) and the clerk of works is lost within him, without any definition of his role if he exists. All roles will therefore be a matter of definition for each project or authority, but clarification should be sought. The ICE form provides for an engineer's representative to be on site and the engineer may delegate to him as he wishes. This produces much the same situation as under GC/Works/1.

THE PROJECT MANAGER

The persons so far considered are those most commonly present and the traditional function of the architect as leader of the team has been mentioned. The growing complexity of many projects is now calling for the exercise of a variety of functions beyond those of design, communication and inspection that the architect mainly performs. Sometimes the predominance of engineering aspects will lead to a consultant having overall control of a project, while a financial emphasis may call the quantity surveyor to the fore. In some quarters the view and the practice is that the total process is sufficiently complex for some person standing aside from the various component functions to undertake co-ordination of the whole.

Then a project manager may be called for. This is particularly the case where a large client has a continuous programme of work embracing numerous contracts in various stages at any one time.

There are two aspects to the work of such a project manager. One is that of seeing the single project through from beginning to end and this may cover evolving the brief, budgetary control, clearing legal and administrative hurdles, obtaining land and finances, while seeing that work of designers, quantity surveyor and contractor is correctly phased, that nominated sub-contracts and any direct contracts are initiated in due time and that the client's commissioning activities run smoothly. The other aspect is that of relating the various projects to the client's total programme and resources. This is specific to the large organisation and will carry the responsibilities involved in the individual project up to an aggregate level. It will also involve advancing or delaying individual projects or switching resources between them to suit progress of the whole. What may be the best timing or emphasis on the single job may make demands that distort the total programme and so the need is for someone who can take this larger view. He will still need to monitor key elements and stages on the various jobs, and this perhaps will mean that he will draw information directly from the specialists involved in each case, rather than from the team leader and will then piece it together as aggregate measures of progress, expenditure or the like.

In all this activity the project manager, whether he is a distinct person or one of the team members with an extra hat, is acting as an extension of the client to make operations more efficient for his benefit. He must not inter-pose himself in the normal working relationship of the architect and the con-tractor during construction, in view of his position in terms of Figures 4(ii) and 6(i). Thus the architect is still 'the Architect' under, for instance, the JCT form and only he may instruct the contractor under the contract. If the project manager wishes to introduce say a postponement of work then he, like the client, must have the architect issue the instruction. Similarly the existence of a client's project manager does not affect the contractor's usual right and responsibility to manage construction itself to produce the end-product of the contract.

STATUTORY OFFICERS

There are various persons hovering around the periphery of the building operation that takes place on a single site. These are the officers responsible for ensuring that the public interest is being observed. To say that they are the client's 'advisers' is rather to understate their powers, but it is primarily at his end of the chain that their effects are felt. In the main they will deal with the architect as the client's chief professional adviser. This is usually clear-cut so far as matters of planning are concerned

Where there is a question arising out of the Building Regulations or some byelaw, a split of dealings is possible. In all preliminary matters the architect is clearly the person with whom the district surveyor or the building inspector, as the case may be, will deal. When the contract is under way, it is the responsibility of the contractor to comply with the regulations in all ways. In so far as this means producing the works designed by the architect there is no problem. Where some discrepancy comes to light between regulations and design, it may be that an instruction is given to the contractor in the first place. The JCT form is the most explicit here in requiring the contractor to seek confirmation from the architect before proceeding, (see *BCPG,* Chapter 5 (a)) although this may be inferred as the requirement of the other standard forms, to the degree that works may be subject to control. The purpose of advising the architect is to allow him the option of a plain acceptance of the position, or of securing a waiver, or of redesigning the section concerned to overcome the problem in a cheaper or more advantageous way. For the quantity surveyor, the main point at issue is to obtain properly authenticated confirmations on all such matters.

The construction organisation
THE CONTRACTOR

As has been stressed, the contractor is the only other party to the building contract with client. Even this relationship is achieved only when a contract has been brought into existence; until then he is the tenderer or one of a list, or perhaps the firm with whom the client is treating. In entering into the contract, he is offering to provide at least three things of major consequence, one being the physical consideration of construction.

Issuing from this, the contractor secondly provides the overall coordination of the many firms who contribute to producing the buildings and other works of today. This management function saves the situation in which the client would be entering into a series of separate contracts with all the firms and the architect would be trying to coordinate their activities on site. Such an arrangement of separate trades contracts is still used occasionally in Scotland, but it is almost extinct. There are however contractors who will enter into a management contract on a fee basis under which they co-ordinate the various firms who perform the works under contract to the client and who are paid by him. The efficiency of the management provided and paid for is related to a target cost arrangement agreed in advance. There are projects of peculiar size or complexity where it becomes rational to engage a number of direct contractors, but where this can be avoided so much the better. The third element in the contractor's contribution is that of providing the funds needed to finance construction, to the degree that they are not made available to him under interim payments ahead of his actual expenditure. Of this aspect more is said later in this chapter.

In matters of procedure, the contractor is again in a unique position in the left to right relationship, since he is the only member of the construction organisation intended to have direct dealings with the architect, and thus with the rest of the design team, so far at least as instructions go. This point has already been outlined; all that need be added is that where any sub-contractor is offering a design service, or the contractor himself is doing so, then that firm may work directly with the architect or with a consultant in the development and approval of the design. At this stage the firm will be acting outside the sub-contract or contract which is to be based upon the design. Even here the distinction must be maintained between a scheme design upon which a price is to be based and the approval of working details already covered by an accepted price. In the second case dealings must follow the procedural network shown in Figure 4, since they are within the contractual arrangement.

But the contractor also has a direct line of contact with the quantity surveyor, because the contract provision if it is explicit is for the latter to prepare the final account. This is to be available to the contractor at some stage of proceedings, so that he may comment on it. On the other flank of his dealings, he is the only person entitled to have commercial dealings with the range of suppliers and sub-contractors, whether they be nominated or not. This is true of any negotiations that he may wish to conduct over site facilities and affairs that may lead to a contra-account to set against the original quotation. It is also true, however, of the settlement of a nominated account in its entirety if the contractor wishes to have it that way. He may deal with the surveyor or the architect on the one hand and the nominated firm on the other, perhaps reaching a different agreement with them on some matters. Usually the contractor will be only too pleased for the others to agree the accounts with the nominated firms informally, so that when they are invoiced to the contractor there will be no reference back of items. It will still be necessary to implicate the contractor in this preliminary agreement if there are contra-accounts or disputed items for work on site, or discrepancies over deliveries.

There are two cases in which the contractor's relations will differ in a number of respects: one is that of the all-in service and the other is that of the management service, both of which are discussed later.

SUB-CONTRACTORS AND SUPPLIERS

Very few contractors today carry a full range of the skills needed for the erection of even a fairly modest building; this is due to the lack of continuity that there may be for their employment and to the sheer organisational problems. Where a contractor chooses to use a sub-contractor to carry out work that has been left at his option as to how it is done, the firm employed is variously referred to as 'private', 'domestic' or 'approved' sub-contractor, the last term stemming from the contractual requirement for such a firm to be approved by the architect before a sub-letting may take place. (See *BCPG*, Chapters 14 and 20.) Otherwise such a firm *is* the contractor and neither architect nor surveyor may deal directly with them and, in particular, the surveyor has no right of access to or concern with, their financial arrangement with the contractor; the contract prices will stand. Where a nomination of a firm occurs as the result of a prime cost sum, the contractual position is the same but there is a right to agree what price shall be paid to the firm through the contractor, within the limits set out under the preceding heading. The standard forms of contract are quite detailed over matters surrounding the process of nomination and subsequent dealings (see *BCPG*, Chapters 15 to 18, 27 and 28).

The term 'supplier' is usually restricted to firms providing and delivering materials or components to the site but not undertaking any work on site. Here again they may be engaged by the contractor on his own initiative or they may come by a process of nomination. It is also common for suppliers, and to a lesser extent sub-contractors, to be specified by name in the contract documents so that the contractor's choice is limited to a single firm or one of a group of firms. Such suppliers may be the direct manufacturers or merely distributors or agents.

Outside the orbit of the contractor's organisation, but operating on site, there may be others termed as artists, tradesmen or specialists. These are persons in a direct contractual relationship with the client and present on site under a contract provision specially permitting it. Unless there are special requirements to the contrary accepted by the contractor under his contract, the latter will have no responsibility towards these persons, other than to allow them reasonable working conditions (see *BCPG*, Chapter 14). Co-ordination is the client's job or better his architect or project manager.

THE CONTRACTOR'S COMMERCIAL ACTIVITY

Two aspects of the contractor's activity may be outlined as being closely related to that of the quantity surveyor. These are the one concerned with the obtaining of business and the one concerned with accounting for it and endeavouring to realise its profitability in the process. Figure 5 shows the main flow lines that are included in the two aspects, with the underlining indicating the main events in historical sequence for a project, whether dealt with by competition or by negotiation.

The decision to respond to a particular enquiry by instigating the action of estimating is one for management to decide in terms of the firm's policy. This may be influenced by the amount of work in hand, the nature of the potential job or its client, or the economic climate, among other things. Contracting firms vary in size, geographical coverage and type of work undertaken and act accordingly. In addition, they are influenced by their own internal information; here the right hand part of the figure shows several influences feeding up information through the accountant. These are looked at below. The same influences also affect the estimator.

It is sometimes said that quantities are a matter of fact, an estimate is a matter of opinion, but a tender is a matter of policy. How far all quantity surveyors would agree with the first of these contentions is open to speculation. So far as the second goes, the estimator in the better organised firms today has an increasing amount of fact to integrate with opinion. Some of this comes directly back from sites, as historical costing data, while other parts of it come in a processed form from the accountant. The former group will include outputs and wastages, while the latter will be largely overhead information stated in generalised form.

But in addition to this material culled from the sites, the estimator also has the benefit of the judgement of those who are going to live with the project when he has passed on to other concerns. These are those service sections that contribute to construction, as shown on the left of the figure. Since they are to be concerned in the operations, it is sensible for them also to be concerned in estimating, where their own effort is involved. Thus a programme in quite reasonable detail will be drawn up for the larger job to assess the preliminaries and intensity of effort required, with a network not infrequently being used. Market information from the buyer will assess prices and shortages, while the way in which it is proposed to set up the hoisting and other plant arrangements will also affect the pricing. Support activities from central facilities need to be considered, and whether in the case concerned it will be better to decide on outside fabrication or the like. None of this should be so constrictive as to prevent a change of method later, but at least a feasible approach is incorporated in the tender.

When all this blend of known fact and intelligent opinion has been made, it will be time for the tender as proposed to be put before the directors or whoever is to make the policy decision that again comes forward at this stage. The sum arrived at will be either inclusive of a stated level of intended profit or entirely net of profit. It will now be the responsibility of management to decide at what level to put in the tender. The securing of

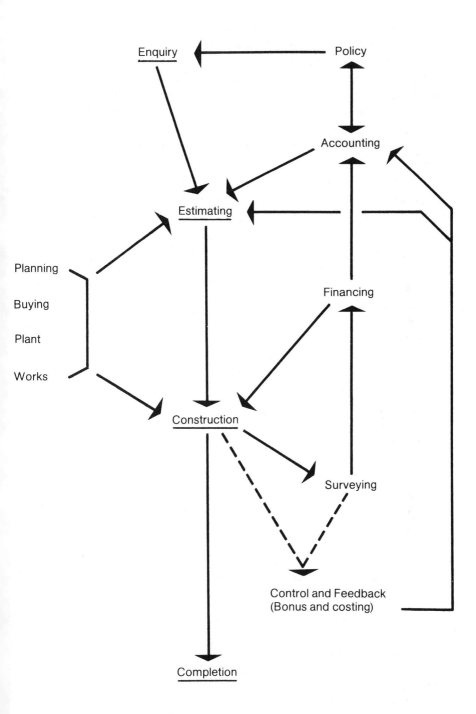

Figure 5: Contractor's commercial flow pattern

some other contract or its loss may affect the desirability of the present project at the moment of decision, as may other changes in the total situation that are beyond the terms of reference of the estimator.

In performing the foregoing activities, the organisation relies on the fact-finding actions of those dealing with financial and other data on site. This aspect is essentially a matter of 'feed-back' and is partly a costing function. It is also a matter of the final account being provided in such a way that the costs incurred can be reconciled with the payments received. However, before anything can be fed back the relevant events must have taken place .Thus the costing operation is a record of the construction process itself; it will give a statement of the actual costs and, it is to be hoped, provide some of the explanations as to why they were as they were, including the success of the site bonusing system. This, it may be noted, is mainly a function of management on site, to secure progress and productivity, even though it has such a financial content. The various aspects are closely interrelated and this outline is here passing over to the realisation of profitability during construction.

It is the duty of the contractor's surveyor to see that profit does not melt away through a lack of proper accounting for all variations and the like. He is operating on several fronts at once in this struggle. He clearly keeps closely in touch with the quantity surveyor acting for the client, but he also has to deal with sub-contractors and his own firm's construction site organisation. It may be seen that there are three main directions in which a contractor stands to gain or lose on a contract to a degree that he did not anticipate:

(a) The tender may be too high or too low in relation to normal efficiency.

(b) The site organisation also may not match normal efficiency.

(c) The surveyor may achieve a reimbursement that runs above or below the level that the other two factors would have suggested.

These factors may of course be overlaid by such things as abnormal weather conditions that are quite beyond the control of the contractor. In looking back on a completed contract, these factors between them tend to receive the most consideration in assessing the outcome.

There remains one other aspect which is of great importance to the contractor, but which it is easier for the quantity surveyor, in his pre-occupation with the final account,to forget, and that is cash flow. No contractor can run his business without sufficient finances to meet his commitments as they arise. He will seek to employ his capital in the best way and the less he needs to have tied up in the period of slack between paying out to others and himself being paid, the better. The more efficiently he can spread these limited resources the greater will be the turnover that he can achieve, and therefore the greater will be the opportunity for securing a profit. A number of his expenses will be in the nature of overheads and will be incurred at times not closely related to expenditure on an individual site. These will include head office expenses, much general administration throughout his field of operation and such items as his own plant, so far as depreciation and other overall aspects go. Some of these expenses will be running items, while others will be large occasional charges, perhaps on an annual basis. In addition to these more general items there are many that are directly incurred due to activities on a single site. These tend to be related therefore quite closely to the rate at which the work done on site is being paid for through the interim certificates, as well as to the final payment made. This is why the contractor's surveyor is more concerned about the sufficiency of these payments than his opposite number may be. Some of the main items fall due in this broad pattern:

(a) Labour: the work force must be paid each week or there will be consequences that are easily imagined. This will be as much the case with any self-employed men.

(b) Private sub-contractors: much will depend on the individual agreement and the substance of the firm. Usually it is possible for them to be paid just after the contractor himself is paid.

(c) Nominated sub-contractors: if the client holds out until the last permitted date in making payments under the JCT form then, even if the contractor does the same, he will find himself paying out before he is paid himself, unless he manages to present the certificate on the same day as it is issued. The other standard forms are not so explicit on the point.

(d) All suppliers: the usual terms are for payment to be due within thirty days of the end of the month of delivery. In a majority of cases this will means that the contractor will be paid first, depending upon the relative timing of certificates and deliveries, as discussed early in Chapter 16. This arrangement is a key source of finance for a contractor.

(e) Plant: hired items often fall to be treated like private sub-contractors. The running expenses of the contractor's own plant will usually be a blend of labour and supplier items.

In terms of Figure 5 it is a matter of the surveyor seeing to it that, as construction proceeds and expenditure drains the contractor's financial header tank, so enough money is brought in and pumped up to the tank to ensure that the pressure is sufficient to keep work going. Otherwise the contractor may find himself insolvent and thus unable to keep in business. This can well happen even if the outcome of the work is going to be quite profitable, simply because an insufficient amount of finance is available at a precise moment and no further overdraft is possible. It is thus a responsibility of the client's quantity surveyor to see that the contractor is adequately paid, while safeguarding the interests of his client. Similar considerations have led to the progressive reduction in retention levels over the last few years (see Chapters 16 and 24 below and *BCPG*, Chapter 11 on 'Payments and Settlements').

Alternative patterns

In several places in this chapter and later reference is made to relating members of the design and construction team to one another, in ways which are alternative to that which has become traditional in past decades. Chapter 9 is concerned completely with one of these alternatives. The salient features are given together briefly here, with gross simplification, so far as they provide the context for the work of the quantity surveyor.

(a) *The traditional pattern*: the architect assumes overall control of design (with any other consultants) and cost (advised by the quantity surveyor) for the client and guides the contractor over implementing the design. (See Figure 4(ii).)

(b) *The contractor in the team pattern*: the structure is the traditional one, but the contractor advises to some variable extent on practical aspects of the design. This may be only pre-contract, or even only pre-tender, but may continue later. It may result in the contractor receiving some form of fee, or sharing in the financial benefits of any design improvements. In formal procedural terms, it is however purely advisory and leaves the main pattern traditional. (See Figure 4(ii).)

(c) *The project manager pattern:* the project manager assumes overall co-ordinating control of the architect, consultants, quantity surveyor and contractor, without changing their procedures and products and without becoming responsible for their individual activities, in terms of either quality or progress. He is however responsible for achieving more effective interaction between these persons and for various client-related activities to obtain the completed project efficiently. As a result he is superimposed on the structure shown in Figure 4(ii), while the client links to him directly rather than to the architect. (See Figure 6(i).)

(d) *The management contractor pattern:* the management contractor is responsible to the client for organising construction, providing central site facilities and engaging and paying contractors for the various elements of the construction work. As an integral part of this, he is responsible for the budget and programme aspects. As a result and according to the precise form of his contract and financial liability, he

(i) Project Management

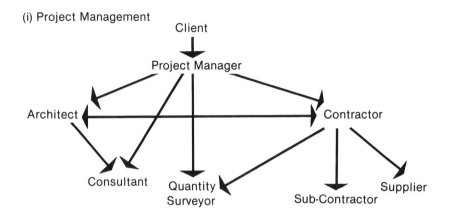

(ii) Management Contracting

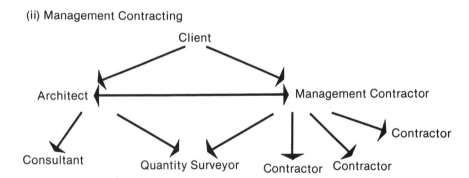

(iii) All-in Service

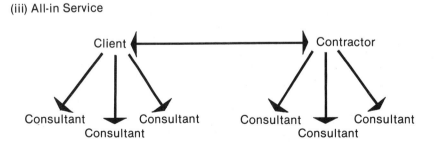

Figure 6: Some alternative design and construction procedural relationships

may take over some functions of the quantity surveyor under the traditional pattern and, while not taking any design responsibility, will perform some co-ordinating of design with construction. The structure will vary here, but essentially the client moves from his position to one side in the traditional pattern to one in which he relates directly to both architect and contractor, whose roles have been modified. (See Figure 6(ii).)

(e) *The all-in service pattern*: the contractor assumes all functions of design and construction in contract with the client. Each of them may choose to retain their own consultants, who act in relation to the other party only if there is a specific delegation recognised by that other party. (See Figure 6(iii).)

A shifting emphasis in these patterns may be noted. In patterns (a), (b) and (c) the architect, other consultants, quantity surveyor and contractor all perform the same primary functions. In fact pattern (b) differs procedurally from pattern (a) only in one possibly transitory matter. Pattern (c) does not change roles and relationships, other than to extract the responsibility for co-ordinating them all and place it with a fresh person. Even so, this person does not enter into the contractor's detailed organisation of construction. Pattern (d) does affect the various persons, in that the management contractor has consultant and executive functions on behalf of the client. He thus encroaches on the roles of the other consultants, while still dealing directly with matters of construction. Pattern (e) carries affairs to the ultimate, as the contractor absorbs everyone except the client.

It is possible to mix these patterns to some extent. Too much ingenuity may lead to unforeseen problems but, in view of the value already discovered in each pattern in given situations, further fruitful development is to be expected.

Except in dealing with pattern (e) in Chapter 9, this book assumes the traditional pattern (a) for simplicity of presentation. All the basic processes of the independent quantity surveyor are transferable between patterns (a) to (d), with relatively minor adaptation. Explicit connections have not usually been drawn, certainly not in detail, between the patterns and the economic, tendering and contractual considerations which are kept apart for analysis in Chapters 5 to 7. This type of clinical separation of course is not possible for any live project, and the reader should draw his own connections in reviewing the various chapters.

PART 2

THE PRE-TENDER STAGE

CHAPTER 5

Introductory economic considerations

The aim of this chapter is to set out a number of economic aspects that underlie the practice of quantity surveying and which are then treated as implicit in the discussion of procedures and methods that follow. It is all too easy to consider procedures and the like without reference to the philosophies that they assume and the impact they produce. The relevance of the points made here should therefore be borne in mind later in the discussion.

Obtaining a building, like any other capital investment, involves a series of decisions that continually narrow the available options within the pre-existing constraints, until there results a specific building to be paid for, occupied, maintained, perhaps altered and eventually demolished. In general, the earlier decisions are the most critical on account of this narrowing effect and, as later decisions are taken, become harder and more expensive to change. Once construction starts, this situation is heightened and hence it is most important to perform the various steps in the pre-tender stage carefully.

The commercial framework for the client

Someone who wants a building has several options open to him. He may buy one or build one; if he chooses to build, he may do it himself, perhaps with the aid of those whom he gathers around him on a paid or voluntary basis, or he may enter into a contract for another to do the work. If he chooses a contract to build, this may be one that includes the design function or this may be a separate element, probably performed by yet another person under contract. If there is a contract to build, there is the choice of having a price or at least its basis agreed in advance, or alternatively of being so rushed or foolhardy as to enter into a contract without any agreement about price. Assuming that a price is to be set up in advance, then this may result from some form of competition or from a single offer, perhaps using negotiation. This outline covers most of the commercial choices that exist and much pre-contract activity is directed towards filling them out in the most desirable way for the particular job in question. It will be seen that the choices fit the pattern of narrowing the options and that succeeding chapters are given over to expounding most of them.

In addition to these choices, there are a number of matters that can be viewed as subsidiary as points of commercial policy, even if more prominent from other angles. One set is concerned with the siting and design of the building, and those aspects that tie closely with quantity surveying practices are considered in this chapter.

Another set is concerned with various matters of timing. The state of the market, in terms of how full contractors' order books are, can affect the cost of the project quite noticeably, and this may persuade some clients to hold off for a while. If they are ready in time they may even launch forth early. There is also a seasonal effect, since certain stages of building are best carried out in the better weather and are therefore cheaper at those times. However any recommendation to delay must be set against inflationary trends and the need to obtain the building by a particular date.

Timing also means starting the whole pre-design, design and construction process early enough to meet the finishing date desired. For any project there is a 'most economical' period for this. This is not a simple issue. To mention only two considerations: a building may be designed to be quicker but more expensive to build, while another may be constructed in a time that gives the lowest final account but extends the construction period so far as to incur extra interest charges on the money paid out, so that these then exceed the saving on the direct payments to the contractor. A policy should be worked out in advance with the client, and this may include some overlapping of the design and construction processes as is considered later in this chapter. It is usually the case, however, that attempts to speed up part way through, to secure earlier completion or to regain lost time, are expensive compared with prior decisions to go at speed.

Comparative cost advice for the client

For the purposes of this section, the commercial decisions just outlined will be ignored and attention will be restricted to the physical aspects of a project in so far as they affect its cost. It is a variable matter as to how soon the quantity surveyor becomes involved in discussions; increasingly clients are calling him in earlier, sometimes before any other advisers.

THE CLIENT'S INITIAL DECISION

When the client is thinking of a new building, his decision whether to build or not is influenced by numerous inter-related factors. Should the quantity surveyor be called in at the feasibility stage to advise there are many of these factors, such as the purpose of the building and the source of finance, that he will usually have to take as given. They are constraints as far as he is concerned. He may well be called in on the next tier of initial decisions, alongside the architect, the valuer and other advisers.

For while the client will usually know the purpose of his building at least fairly closely, he may still be uncertain on the size, disposition of the parts and the character of the building. If he has a particular site already this will condition the solutions that can be obtained. If however he does not have a site, then the question of where to build needs to be considered along with other matters. The economics of a building development are related to the sum of the initial and continuing costs of all types, set against the return on the development and the proceeds of disposal, whenever this is to occur. The return may be financial, but may be notional in terms of evaluating a service provided, perhaps to the community at large. Where the building is situated can affect this return at least as much as can the cost of the building.

The client should be able to answer the locational question if it is as wide as which part of the country, since this is related to the purpose and such issues as the market area. Once matters focus on one locality, the valuer comes into play. Once matters focus on a specific site or sites, the architect and the quantity surveyor can advise on the broad potential in terms of accommodation and special constructional costs due to topography, existing buildings on or adjoining the site and other characteristics.

At this level of generality, there may be hardly any indication of layout — simply an indication of the scale of provision. The quantity surveyor may need to work from either end of the problem. He may be asked for the likely cost of the accommodation at a suitable standard, so that the client can add it to his other figures to see whether the project is feasible. Alternatively he may need to start from the desired return on the investment and work backwards to see whether it is feasible to provide enough suitable accommodation out of the part of the developer's budget that can be allocated to this element of expense. Any advice given should be suitably qualified as to its assumptions and uncertainties, not to avoid responsibility but so that the client and his advisers know the extent of the imponderables and make decisions in the light of them. They can always then seek more defined advice from the quantity surveyor by asking him to assume certain conditions.

When there are several sites under consideration this exercise can be quite complex, owing to the number of variables in play at the same time. Even over one site, perhaps itself one of several, there are questions of possible variations in the plot ratio or other measure of density of development which affect the value of the site to the client or, conversely, the level of his return. If the use of the projected building is open to modification or if the proportions of a mixed development can be changed, as between say manufacturing and warehousing or shops and offices, then there will be different levels of return available according to use, letting or selling price. Usually the client will be looking for the best return.

It may be that the client has an existing building which is not ideally suited to his future needs. Here he has the broad options of maintaining it as it is, altering and renovating it, demolishing it and redeveloping the site, or selling up and starting somewhere else. These decisions introduce a number of changes of relative detail, but in principle bring in similar considerations to those already mentioned.

When the choice of site and the method of development are clear, it will be possible to proceed to the stage considered next, although in cases of particular complexity that stage may be started before some of the initial alternatives can be entirely eliminated.

THE DESIGN STAGE

Only when a site is chosen or closely in prospect will design usually start in earnest. Before it can go any distance, the client's brief must be clear. A sophisticated client will already have seen to this at least in outline, while the sort of financial sifting considered under the last heading will have isolated a number of issues and perhaps have established a firm budget. If necessary the architect will prepare several outline schemes to test the client's reactions over function and aesthetics. These the quantity surveyor should rank in order of cost and then either confirm, modify or set up the budget figure on the preferred alternative for the benefit of the client and the design team. If the client has started with his budget figure or with some cost limit based on accommodation (as often happens with public projects), the order of events may be different, with the design being adjusted to the money. In any case, it is best for the various aspects to develop as the result of continuous close consultation.

While an initial cost plan may be instituted at the feasibility stage, a close one should be formulated as the main design decisions are taken over layout, structure, finish and services. This should be based primarily upon available cost analyses of other projects, with a strong preference for those for which the quantity surveyor knows the historical peculiarities and cost distortions, but otherwise using approximate quantities techniques to the extent that information is lacking or suspect. If this cost plan throws up any marked deviation from the budget, be it up or down, warning should be given. The architect may wish to adjust the design or the client may wish to adjust his requirements. Otherwise the budget itself may be changed, or some compromise may be adopted.

As the design moves into greater detail and hopefully towards finality, cost checks should be carried out to see that particular cost elements are not being distorted, or that, if they are, the resulting redistribution within the budget is acceptable. As the bills of quantities, if any, are prepared they too will act as a further mechanism for refinement.

The techniques referred to under this heading are all within the established armoury of quantity surveying and deal with progressively smaller issues. As has been pointed out, big early decisions are the more critical and the harder to reverse or cancel out by later smaller decisions. This is particularly true over selecting the right site and layout, but then extends down through the range of further decisions. Some points other than matters of design must now be considered.

The influence of the contractor on cost

While many of the commercial aspects referred to earlier crop up in later chapters, there are some facets of the contractor's role that can usefully be underlined here. These facets affect the cost of a particular project and are the result of practice decisions. They are however in that category of matters whose precise effect is not easily quantified in an individual case.

TIMING AND THE CONTRACTOR

In describing the ways in which the client may be advised, it has been assumed that design proceeds within the traditional framework of procedures used for the last hundred years and more. Under this heading the same assumption is made and any departures occur under the next heading. The procedure is enshrined in the very structure of the standard forms of contract, as well as in the pattern of professional and contracting firms that runs through the British construction industry. This procedure dictates that any work meriting a design of consequence, and intended to be built under a contract with the client, shall be designed by one firm or individual commissioned by the client and shall then be constructed by another. Unlike other major industries, the construction industry is in the main orientated to this procedure, both for building and for civil engineering. More particularly, in the case of building the industry has gone one stage further and, by producing the quantity surveyor, has separated the control of cost from the other design aspects, such as function, aesthetics, and structure.

This separation of the functionaries for design and construction often carried with it another separation, and certainly the JCT contract forms with firm quantities and without quantities assume so. This is for design, tendering and construction to be distinct processes in time, following closely but not overlapping each other. It is a constant cause of self-examination by the industry that these clear stages are not achieved. Sometimes this failure is the result of inefficiency, sometimes it is due to pressure from the client. Many clients simply want the shortest time from initial brief to completion on site, so that their investment can start to bring in its return; that the project has cost somewhat more initially is of less consequence to such clients. For others, the interest charges during construction, when borrowed money is doing nothing, may be so high that the client will settle for the shorter construction time and perhaps will even see that this is achieved by adequate planning in what may well be a longer pre-construction period.

This raises the question of when to appoint the contractor so that the traditional procedure may be telescoped if needs be. The traditional answer is — at the end of the design period and after receiving an acceptable tender based on a lump sum without quantities for the small job and with bills of firm quantities for the rest. The process does impose a certain discipline on the design team by the introduction of a crucial intermediate programme date, even though this discipline is not always fully accepted. It is also a key point in enabling the client to assess his financial commitment on construction before he actually enters into it; the contractor's prices replace the quantity surveyor's forecasts. How much more significant the tender figure is at this stage will depend again on how the design discipline has been accepted up to that point, and also on how few changes are subsequently introduced. From the contractor's viewpoint, the traditional procedure is one which gives him, from the beginning, a fair picture of what is involved. Most contractors prefer this, although some would rather have more room to manoeuvre at final settlement; this may be understandable but is hardly in the interests of efficiency or economy. A good contract will always be as well defined as is practicable in all the circumstances.

The answer at the other end of the scale is to start to treat with the contractor, if not to appoint him irrevocably, at the beginning of the design stage. This can mean at the stage of schematic approval by the client or, even more radically, at the briefing stage. As early involvement as this is of questionable value to client and contractor, except when the issues raised under the next heading are relevant. For present purposes, the main aim is secure an overlap of tendering and perhaps construction with design. For this some intermediate time of appointment is likely to serve quite well.

From the quantity surveyor's point of view, the aim will be to permit competitive tendering on reasonable documents or a basis of negotiation that can be developed to an acceptable stage of precision before there is firm commitment. Both of these themes are taken up later in this part of the book. But the client may not appreciate or be influenced by these arguments. His overall view of economy may be to press on with the maximum of haste, with an inadequate contract basis and incurring costly errors all through to completion. If the surveyor finds himself under this pressure, he should advise the client of the dangers of going ahead in this way, while making the best of the situation in setting up the contract. Provided the client has been advised, the decision is his. The contractor should be clear from the beginning as to how the land lies and his price level will then reflect what is known. The rest he can expect to pick up on the way through the contract — and perhaps a bit more as well.

Today's methods emphasize speed more than ever before, but it is also the case that contracts are getting bigger and thus tending to take longer despite all the techniques available to counter this. The desire to speed things up is therefore partly an attempt to do more in the same time, rather than simply be faster as such. The two are opposite sides of the same coin.

Figure 7 illustrates a few of the alternatives and relates each of them to a common design programme. For 'competition' and 'negotiation' both a firm and an approximate basis are shown, giving commitment to a contractor respectively after or before design completion. By 'co-operation' is meant the contractor's co-operation in design work discussed under the next heading and an approximate basis is shown to give an early start.

The practical application of these alternatives needs careful thought but they themselves need thought in terms of which to advise in a particular case. There tend to be fashions in such matters, as elsewhere; merely to call a thing traditional or novel, conservative or progressive, is not to resolve its value one way or the other. The economic considerations given in this chapter are usually what determine value in the quantity surveyor's earthy realm and they should influence his part of the advice the client receives.

When, therefore, the quantity surveyor is able to influence basic commercial decisions he will need to weigh the alternatives carefully. The traditional procedure gives the quantity surveyor a predictable way of working from job to job and thus appeals to his methodical approach. But other things may override tradition in their importance to the client and the quantity surveyor must evaluate these in making his recommendations. In some of these cases approximate quantities may give a suitable early basis, perhaps competitive, so long as it is realised that if they are too far out they may lead to pricing which in the end also turns out to be approximate!

These considerations are not always clear-cut and the client may not want one to prevail to the exclusion of all else; he may also easily be misled in what he wants and may need enlightenment—if he will accept it. The desire to 'get a start on site' is notorious, yet sometimes it has its advantages. Some contracts nowadays are so large that complete prior design is not practicable, while some depend on decisions that cannot be made until half way through construction. Then there is the telling philosophy that the only way to get some clients and designers to make up their minds is to dig a big hole and then ask what is to be put in it! All these things, and more, can affect what the quantity surveyor may recommend about tendering and the contract generally.

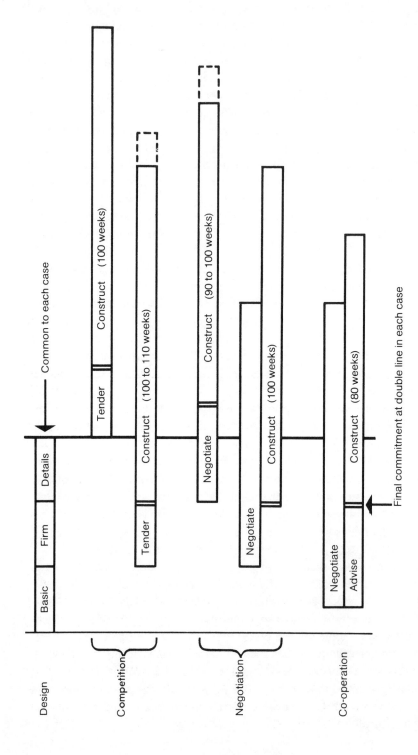

Figure 7: Patterns of contractor selection. (Only a few of the possible variants are shown, with feasible timescales. Proportions of elements are not to a constant scale, overall differences are therefore exaggerated.)

DESIGN AND THE CONTRACTOR

The timing questions just discussed have been seen as leading to adjustments in the phasing of design and construction, but not as involving any change in the principles of designing in particular. These are set within the traditional pattern of design by one person and construction by another, which itself evolved from an earlier phase of a master designer who also organised the construction work by engaging groups of men in the various trades and co-ordinating them. In the last century, the role of the general contractor became established and the task of organising passed to him, leaving the design role with the architect.

There has been much discussion over the last few years about the desirability of this division of labour. It does afford a measure of financial and design control which few other industries operating under contract are able to equal; this presumably brings benefits to the client and, indirectly, to the contractors. Whether it contributes to the low profit margins of the industry, and thus to the low level of research and the high level of insolvency, is a matter of conjecture. What is important is that it limits the possibility of feed-back of construction experience occurring and being taken into full account in succeeding designs. There may thus be a situation of efficient control of cost, but with the underlying level of cost too high. It also divides the cost of designing from that of construction, and puts an arbitrary proportionate valuation on each, by virtue of the professional fee structure. If the two were both part of the same firm's cost then presumably the proportions would be optimised to secure overall economy in the end product. Sometimes more design would be warranted, to save money on the site, and vice versa.

In the last decade or two there has been a growing tendency to involve the contractor in design matters. This shift has produced two types of operation, although the argument in favour of both is essentially the same — that is, the contractor acquires considerable practical experience on matters such as sequence of construction and continuity between operations (with resultant effects on plant utilisation and gang balance) of which adequate account cannot be taken in the architect's drawings.

The first type of operation involves the contractor as part of the team, co-operating to suggest ways in which the project could gain from the practical standpoint without losing the character that is otherwise desired. Here sequence and plant questions are the main concern, but in some projects the solution to a structural problem may stem from the special experience of a suitable contractor. This needs either early appointment of the contractor with some price basis set out, or a consultant arrangement supported by a fee or, more dubiously, by the carrot of a nearly certain contract. The aim is clearly to secure economies by the approach. It also allows an overlap between design and construction if there is an early appointment, possibly based on negotiation as considered in the next chapter. On the other hand negotiation does not necessarily give a low price, and it may be a case of getting a better job for the same price, with the suggestions made counterbalancing the market level effect.

The second type of operation is the all-in service in which the contractor assumes the design function as well as the construction function — hence the term 'package contract' often applied to this form. This may well bring the contractor in at the briefing stage, although perhaps not with final commitment until much later. It can involve the architect and other consultants to evolve the brief, or to check the contractor's proposals as they develop. This approach can lead to substantially cheaper building, but then has to be weighed against the quality of the finished product. If proprietary system building is envisaged, the approach becomes almost inevitable. Chapter 9 enlarges on the points at issue.

PRICE AND THE CONTRACTOR

At some point there must come an agreement as to the level of price that is to be paid to the contractor for whatever he is to produce and for whatever responsibilities and risks he is to assume. It will usually be said that this should be a fair price. The term 'fair price' is riddled with ambiguities and is not helped by the lack of a market price for a building (proposed or existing) that is equivalent to that usually held to exist for, say, some common article of food. It is usually assumed that competitive tendering or negotiation in particular market conditions produces a fair price in the sense of market level. The related procedures are covered in the next chapter.

Since buildings and other such works, even at their simplest, are fairly complex assemblies, it is further necessary to define a basis for paying the 'fair price' that may vary in complexity with the construction process, and here the client's and contractor's interests are not in complete harmony. This is illustrated by the following basic diagram of the construction process:

| Contractor's costs (inputs) | → | Construction activity (black box) | → | Work produced (outputs) |

Here the client has a naturally strong interest in paying no more than he need for work of given quantity and quality, produced when he wants it, so that his interests are likely to be best served by a contract in which the commercial terms provide for him to pay a controlled amount for controlled outputs produced. He will not be concerned about the 'black box' (in the systems sense of a discrete process, the internal workings of which need not be explored) unless it affects the outputs, and he will have even less interest in the inputs. If the contractor is to be paid in this way however, construction activity will be anything but a black box for him since it is here that he must seek to use the minimum set of inputs for a given set of outputs and thus maximise his profit.

Following on this elementary discussion it is not surprising to find that 'payment for work produced' is used as a basis whenever possible. When circumstances of risk or other uncertainty make it necessary to set up a basis of 'payment for costs incurred' it is not surprising that the client's interest widens across all three parts of the diagram, while the contractor will remain vitally concerned with being paid for the first part but will possibly be less concerned with the other two. These simple motivations underlie the discussion in much of what follows. They apply not only in the initial basis, but also in conditions such as increased market costs and unexpected loss and expense that may lead the contractor to seek a measure of cost reimbursement from the client in an otherwise 'payment for work produced' contract.

Standard contracts commonly contain provisions to allow for this hybrid method of payment, but the significance of the change being introduced and the accounting difficulties of having two methods, should not be ignored. In particular, it is only possible to make the change to the extent provided by the conditions — if it is suggested that a complete change to paying 'what the job has cost' should be made so that unexpected conditions can be fully covered, this is changing the fundamental structure of the contract and can be done only when there is agreement by both parties to the contract. It is not a course that professional advisers will lightly recommend in any case.

One approach that has been developed over recent years and that tends to cut across the two bases is that of the management contractor. Here the contractor is appointed under a fee arrangement to organise the production, including engaging firms whom the client pays. There is a target price that the contractor meets or betters and his efficiency is judged, and perhaps his fee adjusted, in the light of performance. The general framework has a ring of prime cost, as discussed in Chapters 7 and 18, but the details may relate to measured or lump sum bases.

A summary

A number of factors have been touched on quite sketchily in this chapter, essentially to highlight the variety of matters that affect the cost of a project to the client. The term 'factors' is appropriate in that these matters do come together in a multiplicative way. Thus it could be said that an easy or a difficult site can alter the cost of a building by a percentage from some assumed average price on the non-existent average site. To say this is to take account solely of the cost of producing the building and to ignore altogether its suitability and thus its value — another factor of importance. It is considerations of these types that are usually classed under the umbrella of cost advice as far as the surveyor's activities go. Concern focuses on the relative merits of one design and siting against another, and this produces a set of differentials that retain their validity.

While this is so, the whole set of figures can be moved up or down by a factor for two broad groups of reasons. One is matters of economic climate over which there is very little control so far as the individual project goes — it is either built now or postponed. The other group consists of matters that fall within the field of practice and some of these have been touched on in this chapter. They form the substance of all the following chapters. Commercial decisions can gain or lose money as surely as can design decisions.

CHAPTER 6

Procedures for obtaining tenders

It is common and useful to distinguish between obtaining tenders and the closely connected subject of the contract arrangements. These arrangements form the subject of the next chapter and the term covers the basis of payment and other conditions embodied in the contract, such as bills of quantities or prime cost.

Tendering procedures cover the various methods that may be used by the client and his advisers to obtain offers. The offers are made by the firms approached, who do the actual tendering and have their own procedures. But to arrive at this point there must already have been a decision on the contract arrangement that will be entered into on acceptance of a tender. Thus the subject of this chapter and the next are inseparably linked in practice but may be distinguished in concept.

Open tendering

The procedure here is to advertise through the press, inviting any firm that wishes to do so to submit a tender; the advertisement will give outline details of the type of work, its scale, the programme, and any other key features. Interested firms apply for the tender documents and there are usually no formalities other than a deposit of a few pounds, which is returned on submission of a bona fide tender. The deposit covers the cost of the documents and discourages idle curiosity. This system is still used fairly often by local authorities, though it is far from preponderating. With suitable encouragement from central government its use has been declining rapidly; government departments themselves ceased to use it just before the last war, and it is a rarity in other circles.

Open tendering does have a number of clear advantages. It gives the chance of tendering to any firm that wishes to do so and may provide the longed-for opening for some relatively unknown but capable firm. Since there is no restriction there can be no charge of favouritism in drawing up the list of tenderers; this is a valid point in the case of public accountability. Just as there is no restriction on who may tender, so there is no obligation to tender. Presumably no one is going to put a tender in just to oblige, and therefore all tenders should be genuine. They should also be genuine in another sense, since it effectively prevents firms in an area from forming an exclusive ring and keeping up prices. But above all else, the appeal of open tendering is that it should secure maximum competition.

The various government-sponsored reports on the building industry which have been made since the war have come out strongly against open tendering; the Banwell Report of 1964 was no exception to this. The strongest objections are over its effect on the cost of tendering and its unsuitability as a means of selecting a capable firm. The former objection arises out of the number of tendering firms likely to be attracted. Only one can succeed for each contract; the higher the proportion of unsuccessful tenders the greater the cost to be recovered on the jobs which firms do secure. No client inviting tenders with any regularity can escape this extra cost and, in addition, there is a severe waste of skilled estimating time.

The difficulty of selecting the right tenderer may well affect the final outcome of the project. Any client choosing open tendering (it is usually a committee that does this) needs some convincing before by-passing the lowest tender; to do so is to miss a bargain and also to invite the charge of favouritisim again at that stage. But even if the client is prepared for such advice, it is not always clear and easy to give. It is not so difficult to reject the firm that is known as giving poor value, but a difficulty arises with the unknown firm whose record needs to be checked at a stage when time is pressing. For an investigation to be satisfactory the same criteria need to be applied as are considered under 'Selective tendering' below and these take time. Another difficulty comes with the known satisfactory firm that gives a sub-economic price in the undue keenness of competition; if there is a major and obvious error it will be discovered but this does not necessarily apply to tight margins.

One way or another the wrong firm, capable or incapable, may obtain the contract. All praise is due to the competent firm that gives good work while it is losing money but otherwise the danger of poor work arises and there are only two answers to this—close supervision and embittered relations or the acceptance of a poor job with permanent defects. Even if the quality is satisfactory, the wrong contractor may have poor organisation and may give late completion. The fractious will seek to recoup their loss by claims and hard bargaining on the final account. Probably the worst result of all is when the wrong firm becomes the insolvent firm, dragged down by incompetence or low pricing on its jobs as a whole, and involving its clients in expense and delay.

Many good construction firms avoid open tendering absolutely, while others resort to it only in times of dire need—which, in itself, is probably a good reason against its use, so far as the client is concerned.

Selective tendering

This procedure is to select a limited number of firms and invite them to tender. Selection should be made sufficiently early for firms to be asked firstly whether they would be willing to tender at the required time, subject to the state of their order books. An unduly large list of tenders is to be avoided and the 1977 edition of the Code of Procedure for Single Stage Tendering suggests numbers from five to eight as contract values range up to £1,000,000 and then a drop back to six. This reflects the amount of work involved in tendering and then the limited number of firms likely to be available for the large projects. It is always desirable to have one or two reserve firms who can be asked to tender if any of the selected firms drop out at any stage.

At the initial approach, sufficient description of the project should be given to allow the firms to appraise realistically what is involved and therefore whether they are interested. This should cover the same ground as that in an advertisement inviting open tenders, but in a letter it can be rather fuller.

The criteria to be employed in drawing up a list of tenderers will depend to some degree on the character of the project, as well as on its size; the firm that can undertake a motorway may not be geared to a patchy alterations job. Apart from an investigation of that aspect of a firm, it is usually desirable to look at the following:

(a) The standard of workmanship.
(b) The equipment, such as plant and workshops, carried by the firm, the size of the pay-roll, and the degree to which it is the firm's practice to sub-let parts of the work.
(c) The business record and standards—for example, whether completion dates are met and whether there is any regular difficulty over supervision of quality or final settlements; the board of directors may be worth reviewing in this connection.
(d) The financial stability and length of time in business.
(e) The capacity available in relation to the current work-load.
(f) The local history in respect of labour relations—whether there appear to be more or fewer than the average number of stoppages because of local disputes and whether it is relatively easy or hard to attract labour.
(g) The real willingness to tender.

If necessary, references should be taken up on some of these criteria, while a visit to completed work or work in progress may be needed for others.

It is the practice of many public bodies and others to maintain a standing list of firms that satisfy the appropriate criteria, perhaps sub-dividing that list according to the types of work, the size range of jobs, and the locality.

In the case of public bodies particularly, the list will have been compiled by selection from firms applying to be included in response to a public advertisement. The list should be reviewed periodically in the light of response to further advertisements. When a particular project approaches the tender stage a short list is prepared from the standing list, preferably by taking the eligible names from the list on a rota basis for successive contracts. Any other firm on the list that specially asks may be included as a concession. Any firm approached that does not wish to tender should be excused and should be assured that its future eligibility will not be prejudiced.

Where a formal standing list is not maintained, an *ad hoc* tendering list may be drawn up from the known suitable firms. Alternatively, a similar list may be selected from firms invited by advertisement to submit their names to be considered for inclusion in the list for that particular project. These methods are not so regularised as a standing list but the second one does ensure, like open tendering, that only those who have shown genuine interest are put on the final list.

Selective tendering makes good many of the deficiencies of open tendering, especially as it provides a restricted but adequate list of technically suitable and sound firms, of comparable standing, capable of carrying projects through in a reliable manner. It should not overload particular firms, provided that they are frank in declining invitations to tender when their existing work load renders this advisable. The favouritism aspect needs to be watched with care, but the standing list procedure should overcome any qualms. It is possible however that the whole system may come under review as not complying with EEC policies.

One frequent practical weakness with a standing list is a reluctance to take the bad performers off it. To remove a firm is often felt to be more of a smear than not to include it in the first place and so the spectre of victimisation arises. Nevertheless, if the list is to remain manageable and valid such removals must take place if they are justified. The regular review of the list, with public advertisement to invite new candidates for inclusion, is also necessary if good new firms are not to be kept off when they wish to be put on.

It is true that, in most cases, selective tendering will mean higher quotations than are obtainable by open tendering. This is partly because there is less competition, and therefore a greater likelihood of success in securing the job, but also because the standard of firms, and that of their workmanship and performance, will on average be higher. As in other fields, the client gets what he is willing to pay for — at least in most cases, it never being possible to eliminate entirely the chance of a lapse in standard of work on financial failure. But the level of tenders will in part be lowered by the higher efficiency of the firms, which should be reflected in their prices.

There is the possibility of tenders being 'rigged' and inflated by collusion if firms know the probable limits of the tendering list in a particular area. The safeguard is clear; the selected list should be changed, in part, from job to job even if a complete change is not possible each time. But where geographical considerations make it impracticable to bring in outsiders it becomes difficult to break a ring even if the list is changed slightly from time to time. The same thing can apply to open tendering in a sufficiently tight area.

A variant cause of high tendering on a selective list may be that some firms do not wish to secure the contract but may not consider it politic to decline the invitation to tender. As a result they will put in tenders which are either so high as to put them out of the running or sufficiently high to ensure that if they secure the job they will do so on really attractive terms. In either case, these deliberately high tenders restrict the number of firms giving really competitive tenders and thus there is a smaller selection than was intended. The lowest tender will therefore often be higher in such cases than it might otherwise have been. In any case of high tendering, and where further firms are available, there is always the remedy of re-tendering if time permits.

While selective tendering needs care in its application, it is the method which finds most favour in those cases where competition is desired.

Single tendering

There may be circumstances in which only one firm is found to satisfy the criteria for selection. This is more likely in the case of a specialist sub-contractor but it may also arise in the case of a main contractor. If it does, it becomes a special case of selective tendering and should be treated as such. It is a somewhat undesirable and perhaps unhealthy state of affairs and can be embarrassing if an unsatisfactory tender is received. If the procedure is embarked upon the main point to watch will be that the firm concerned is not told, or led to infer, that it is in fact alone in the field. If this cannot be avoided it is far better to settle for a negotiation in the first place, rather than receive a tender that is then going to be haggled over from a position of weakness. In fact the only virtue of a single tender is in those cases where it is felt to be the best way of obtaining value for money.

Negotiation

Certain of the techniques of negotiation are considered in Chapter 13; the present concern is with the broad pattern of negotiating procedures only. These are looked at here for convenience of arrangement, although negotiation is not tendering in the accepted sense of one party making an offer based on its own assessment of the project in isolation from the other party. But negotiation does lead to an offer open for acceptance in its entirety, even though representatives of the two parties have had many preliminary dealings. The two concepts may be distinguished in more detail as follows.

In tendering, one party makes an offer to the other, usually as the result of his own unaided assessment of the commercial aspects, taking into account the information provided as a basis for tendering. Such an offer should be either accepted or rejected by the second party and should not become the subject of a negotiation in normal cases. Acceptance should be subject only to any clarification of intention that is necessary or to any modification of the object of tendering arising since the original information was issued.

In negotiation, the two parties proceed by proposal and counter-proposal to a situation of eventual agreement. The proposals may concern quite limited areas of the total negotiation at any one time and the separate areas may need reassessment as the negotiation as a whole moves forward. The key principle is that, while progressive agreement is reached on particular aspects, neither party is finally committed until the pattern of the whole is established. The process thus achieves a total offer which, like a tender, is then open for acceptance in totality.

BROAD PATTERNS IN NEGOTIATION

The patterns of negotiation are subject to almost infinite variety; they certainly will vary as widely as the economic and other considerations set out in Chapter 5. Three broad patterns may be taken to indicate the range of possibilities that the character of the project, and the client and his finances, may give rise to in conjunction with the time available.

The simplest pattern is the single-stage negotiation. Here the proposals go to and fro between the client and the contractor until complete agreement is reached. Which party makes the original proposals will depend on the way things have developed; there is some psychological advantage in having the first say. On agreement the contract is entered into in the usual way.

An alternative pattern is that of the two-stage negotiation. In this case a limited selective competition opens the first stage, all firms concerned being given fair warning of what is intended to follow. This competition may aim to reduce the field to one firm only or, at most, to a manageable number of say three or four, the basis of selection being one or more of a number of factors such as:

(a) Leading measured rates or a general basis of pricing.
(b) A construction programme or some other form of method statement.

(c)   A design offer, particularly in the case of system building.

Having obtained what are virtually statements of intent, and having decided which are worth pursuing, the client will carry through the first stage of negotiation to arrive at an acceptable but still broad basis with one firm only. This may mean parallel negotiations with several firms, or successive negotiations if discussions with one firm run into difficulties. At this stage the temptation to develop a 'dutch auction' needs to be resisted. When this stage is complete, all firms but the one chosen should be advised that they are no longer in the running; some indeed may have been told this earlier. The second stage of the negotiation will then proceed until finality is achieved and the contract is awarded. Alternatively, the contract may be awarded at the end of the first stage and matters may proceed as a progressive negotiation.

Progressive negotiation is the third pattern. The principle in this case is that negotiation proceeds to a stage of agreement and the parties enter into a contract at that stage; negotiation is then completed in circumstances which allow no turning back. Shortage of time will be the usual reason for a progressive negotiation, although it may also be considered desirable where aspects of the project cannot be determined in advance or where the economic climate has particular variables. In an extreme case, a contract may be entered into with virtually no agreed basis established. More often a progressive negotiation will be a variant form of a single-stage or two-stage negotiation.

CONSIDERATIONS AFFECTING NEGOTIATION

Negotiation does offer a number of advantages that have led to its being used much more often in recent years. It may sometimes offer a time saving, even when it replaces the normal tendering period only, since the two parties may be able to take short cuts. It should certainly save time when it overlaps the design or construction phases, or both. Also, more factors can be discussed in detail between the parties in the course of a negotiation, including some matters of construction methods and procedure. This should lead to a more rational price basis for the contract and perhaps stimulate design improvements or a reduction in construction time — particularly where designer and contractor are working together over the design itself. But probably the greatest appeal of negotiation is as a means of securing a particular firm to carry out the project; equally, it has an appeal to many of the more businesslike contractors by securing work on better terms of finance and organisation from the client's side. Competition no longer depends on price alone but on reputation also.

Against this is the higher price level that normally accompanies negotiation. This arises partly from the advantages that are being bought but also from the fact the negotiation almost inevitably leads to higher profit margins. But here it is often argued that the average margin in the industry is far too low and a higher profit margin in the initial negotiation may be offset by an easier attitude at final settlement. If the client, or a friend of his, is going to be in the market with more negotiated work there is little point, from the contractor's point of view, in killing the goose.

Another disadvantage, in some people's minds, is that with negotiation it is harder to satisfy the test of public accountability, where this applies. Very often, however, suitable yardsticks can be devised, which act as a check on the negotiated prices, and it is a fact that negotiation has increased considerably in the public sector in recent years, with some of the more advanced techniques being developed there.

In any negotiation several things are desirable if they are at all possible. One is that the essential basis, at least, should be fixed before either party is induced into a contract; where everything is susceptible to reasonable agreement there seems little virtue in leaving things outstanding when time is available. Another is that a 'fair' price, (touched on in Chapter 5) should be the aim. This may not be the same thing as a strictly competitive price, for the reasons considered above, and it will be necessary to decide in advance how far the client is prepared to go. Further, it is desirable that negotiation should not become the sole rule, even if it is not too much of an exception; it should be tested by competition quite frequently. Lastly, if either party puts his professional advisers into action with a brief to succeed at all costs, then it may literally be at all costs. No one can negotiate successfully unless he has the option of breaking off negotiations at a pre-determined point.

Extension

This is often a special case of negotiation, arising when it is desired to place another contract with a firm already engaged in an existing contract, often on the same site. In this case the provisions of the existing contract will be adjusted to suit the new and the prices will be negotiated to take account of these provisions and of rises in prices and other factors.

Extension may be made competitive, however. There may be more than one firm already working on existing contracts of similar type and each may be asked to give their respective terms for the new work in competition, using the old terms as a basis. Again, it may be possible to put the firm which is on the existing contract into competition with fresh firms tendering on new invitations; the difficulty in these cases is to make ready comparisons between the two types of offer received.

The benefit of an extension contract is that it gives continuity of work to an established site organisation and, if properly phased, should utilise supervision, plant and following trades to the best advantage. There will also be the benefit of repetition, and increased speed of working which results. The extent of these benefits will depend on how similar the two contracts are; the result of the benefits should be relatively lower prices in the later contract and better profitability for the contractor.

Serial tendering

This is a premeditated form of extension and comes essentially into the category of tendering rather than negotiation. Like extension, it is useful for a series of closely similar schemes which are to be carried out over a period of time within the same area and permitting the phased utilisation of the site organisation and plant. It is useful for main contracts and can also secure the benefits of increased size of order and production run for component contracts and specialist site installation work, which may become sub-contracts to a series of main contracts.

The basis of tendering will be to tender for the first project or so in the series, as a contract or contracts in their own right. Within the contract arrangements will be the provisions forming the basis of succeeding contracts; these may be in the nature of a formula to take account of the factors affecting prices, with lump sums for preliminaries and the like. Alternatively the contract may provide for negotiated adjustments, defined with less rigidity against a broad basis. This may be more realistic than too algebraic an approach determined some time ahead of the actual circumstances. However successive contracts are to be entered into, the initial arrangement should be such that it takes account of the benefits of repetition by arranging that these accrue in the respective contracts in which they occur, and are not spread over the whole series as a flat-rate saving. To do the latter would be to invite claims in the event of an unexpected break in the series.

The clear intention of a serial arrangement should be that both parties carry on through the whole series on the basis laid down, but there may be unexpected developments that make it clearly inequitable to carry on. The original contract should endeavour to set limits to the circumstances for proceeding, as clearly as is reasonable, but since the whole system envisages a series of distinct contracts there is always an option to break off at each stage. In most such arrangements things finally depend on the goodwill of the parties. Where conditions do change, either the series should be broken off, and the remainder made the subject of fresh competition, or a new basis should be negotiated.

CHAPTER 7

Contract arrangements

Let it first be repeated that contract arrangements can be separated from tendering procedures in the mind but that in practice some form of each must be employed in any contract deserving the name. A comparison of the headings in this chapter with those of the preceding one will show that most of the combinations available are at least possible in practice, although some of them are far more reasonable than others. It is also possible for some of the arrangements discussed in this chapter to be combined in the same contract; this point is taken up briefly at the end of the chapter.

The arrangements considered below seek to deal with the payment for building work on one of three bases:

(a) By payment of a defined sum of money for a defined amount of work, with little provision for adjusting the sum according to any pre-determined formula in the event of variations.

(b) By payment on the basis of the contractor's costs, subject to certain of these costs being treated as a form of overhead, and with only a limited control, at best, of what the costs will be.

(c) By payment on a measured basis for the quantity of work done, using prices per unit quantity and a system of standing charges for the project as a whole; this may be so dealt with as to give a definite sum of money for a definite amount of work right from the outset.

The last of these three bases has generally been adopted for works of any size carried out in normal circumstances; from the principles it embodies comes the very term 'quantity surveyor'. Yet this basis has been queried even in normal circumstances—particularly, though not solely, where a job is subject to remeasurement. The query is whether a standard rate for a unit quantity, when applied right across the job, fairly represents the cost of the work as carried out or whether there should not be distinctions between different parts of the job to reflect position and timing, whether by way of sequence or continuity. Some of the attempts to deal with this query, without going outside the framework of measurements and rates, are partly responsible for the forms of bills of quantities considered in the next chapter.

The other two bases of payment provide other answers to this query, although this may not be the direct reason for their adoption. The first basis does it by leaving it entirely in the contractor's hands to calculate his price by whatever method he may choose, and thus to analyse the project in his own way. The second basis does it by letting the costs 'come out in the wash' — or perhaps be concealed in it. The broad discussion of the main difference has been given in the latter part of Chapter 5.

Another aspect of this same question as to the adequacy of measured rates is how far and how easily they can reflect the changed circumstances of the individual items. One such circumstance will be where a contractor participates in design development after being selected on the basis of rates; conditions of working, as well as quantities, may be affected. It is not impossible to make appropriate adjustments but the rates as such will then be of only limited use. But while measurement has its limitations it is very difficult to see any rational system that does not use it in some way or another unless contract figures and cost control are to be ignored.

The most common arrangements for contracting on these three bases are discussed in this chapter—with the exception of the 'all-in service' which, because it calls for a discussion of more than the actual contract arrangement, is dealt with in Chapter 9. The discussion in this chapter covers the advantages and disadvantages in principle of each type of arrangement, along with some more or less detailed points that can be included without undue digression. In the case of bills of quantities, however, there are sufficient points of detail to justify their consideration at length in the next chapter.

The arrangements are here placed in order according to the firmness of the financial commitment that they give at the time of entry into the contract, assuming no variations. The most definite is given first.

Drawings and specification

This arrangement is the traditional 'lump sum' system used for small contracts and also for many sub-contracts to larger contracts. Its virtue is that it avoids a disproportionate amount of pre-tender work on quantities on the client's part in cases where the tenderers can produce their own, in an abbreviated form, without too much labour being wasted by the unsuccessful firms. The tenderers can afford to abbreviate because only they, individually, need to understand what their quantities mean and, in particular, which items they have not measured. The only real disadvantages of the system are that it is restricted to small projects, that it requires complete designs to be ready before tendering takes place, and that it is not intended to provide for extensive variations. Some may consider this last point an advantage.

The documents required are the complete working drawings and details, with a full specification which will include the preliminaries. The specification will include clauses corresponding to the materials and workmanship content of bill preambles, and also schedules or other positional information relating the other clauses to the drawings. PC sums, and even provisional sums, may be included in the specification, although if this is done the boundaries of the work concerned must be defined clearly in the documents so that the tenderers can assess what to price outside the sums concerned.

Since the tenders will each consist simply of a lump sum it should be written in that the tenderer who is likely to be accepted should provide some means of valuing variations. For additions this could be done by daywork terms sent in with the tender, although these terms are not ideal. Being separate from the tender sum they may not be so competitive, while later they will be useless to value a variation that is of the nature of 'extra over' an element of the original contract. Where omissions are concerned, daywork terms are in the nature of things impossible. A second possibility is to append a schedule of rates covering the most likely items and ask all tenderers to price this when tendering; the only snag here is that these prices do not affect the amount of a tender and may therefore be inflated or deflated according to whether tenderers think they are more likely to be applied to additions or to omissions.

A sounder procedure for the larger drawings-and-specification contract is to ask for a breakdown of the tender under consideration. This can consist of several lump sums that will localise the spread of money sufficiently to 'contain' the effect of variations. Alternatively, the tenderer can provide the detail on which he prepared his tender; the prices, but not the quantities, in this will then be used to value variations as discussed in Chapter 12.

Where a project is quite minor it will seldom be worth incorporating apparatus to deal with variations. The two JCT contracts suitable for the larger and smaller types of drawings-and-specification work make a distinction in this respect. (See *BCPG* Chapter 21, section 'Private edition without quantities' and Chapter 25, 'The Agreement for Minor Building Works'.)

Bills of firm quantities
This arrangement has been the traditional pattern for the larger type of contract since the quantity surveyor became an accepted independent member of the chain of building production. The results is still a lump sum, as with the drawings and specification arrangement. As already mentioned, it is the keystone of the JCT form with quantities and it is also one of the common forms under Form GC/Works/1. In all these cases the quantities form part of the contract and the proper procedure is for the design to precede the award of the contract. This divides the building process sharply into a pre-contract and a post-contract phase and usually, though not

necessarily means that different persons undertake design and construction. Certain quality and financial controls also become separate from construction.

Several advantages flow from firm quantities so prepared. One is that the several tenderers are saved the task of each analysing the works as a preliminary to pricing them, since one person has done it for them all. Admittedly the surveyor takes longer over the matter than any one of them might do. This is because the surveyor has to present his measurements in a form intelligible and unambiguous to other people, while the tenderer can produce data entirely for his own consumption. But then the surveyor's bills of quantities also have to be clear as a contract document.

Another advantage is that proper operation of the procedure imposes a quite rigid discipline on the client and the design team to make up their minds before committing themselves. Even when its operation is faulty it probably improves the situation in most cases. It also means that the client and the contractor enter the contract with their commitments clear, variations excepted; this can never be a disadvantage—to put it mildly. Usually it means that the analysis needed to produce firm quantities, by producing all the measured items, will give the best basis for the valuation of variations. The exception will be where variations introduce work of a different character. Some anticipation by way of provisional items is possible without destroying the advantages of the system. In addition, firm quantities, when priced, do throw up the earliest available accurate cost analysis information of an historical nature that a project can provide. On any other measured basis one has to wait and often the information is too dated to be of much value.

The greatest objection usually voiced about the quantity surveyor using firm quantities is the time he takes and the time he forces the designer to take—or at least tries to make him take—before tendering can take place. The larger the project, the more acute this problem becomes and firm quantities cease to be worth while in the overall balance of considerations. In all cases where work must start only partly designed, firm quantities are out of the running and this will include cases where the work is inevitably partly unknown in character, as with some alteration schemes. In cases where firm quantities are unrealistically expected, pressure of time often leads to quantities which are firm in little but name. The numbers of variations that result, and the artifices adopted to make the final account look something other than a straight remeasurement are sufficient to point to the invalidity of the original arrangement. Often enough remeasurement becomes the only tenable course.

A different type of objection to firm quantities comes from those who advocate that contractors should be more widely encouraged to put forward alternative design solutions when tendering. If the alternative is an alternative to the originally developed design then it will amend the quantities, and possibly do so drastically. If contractors are invited to develop partial designs as is the practice in parts of Europe, then initial firm quantities become impossible—at least in the traditional concept of specific items described in detail and measured in accordance with the Standard Method of Measurement.

There are therefore a number of forces that tell against firm quantities in several, but not all, modern situations. There are other forces that may popularise them in the future, with some change in their content and presentation. These arise out of the demand that quantities should perform more functions; this is enlarged on in the next chapter.

Bills of approximate quantities

These are quite a common arrangement in traditional contracts though they are often adopted with a feeling that they are a second best to using firm quantities. They usually arise when time is pressing. Very broadly, the larger a project is the more difficult does it become to allow adequate time for firm designs and firm quantities to precede tendering. In these conditions, approximate quantities at the contract stage and complete remeasurement to achieve the final account may be almost inevitable. The JCT forms include special editions related to approximate quantities (see *BCPG,* Chapter 21), which are intended for cases in which the works have been 'substantially designed' and the quantities and drawings are 'a reasonably accurate forecast'. Some careful adaptation of wording would be needed to make them suitable for less definite cases. Form GC/Works/1 recognises approximate quantities as one of several options (see *BCPG*, Chapter 27), while they are the only method under the ICE form (see *BCPG*, Chapter 28). Neither form attempts to limit the degree of approximation that it permits and problems due to contract wording alone should not be expected.

PRINCIPLES OF THE BILLS

To use such bills it is necessary for some design work to have taken place. The degree of accuracy that is possible is limited in the first place by the accuracy of the information provided and only secondly by the time available. Approximate quantities may be as detailed as firm quantities, and as reliable in their range of items; it may simply be that their accuracy cannot be assumed to be closer than x per cent.

At the other end of the scale approximate quantities may be no more than an informed or even uninformed guess. The question will sometimes be asked: 'How long do you need if the quantities are only approximate?' An answer could be: 'This afternoon to prepare the draft bills; I will just alter the cover and description of works for any other job in the office and leave the quantities as they are!'However, even where information is sketchy the aim should be to achieve the following contents in the bills:

(a) A total tender that is close to the estimate for the project. The estimate itself may be the stumbling block here, or it may be the only source of information, giving purely element totals. The bill quantities may then need to be roughly priced if their overall quantity cannot be checked in any other way.

(b) A fair balance in the 'weighting' of each trade or work section should be struck. Again, rough pricing may be used if other information is scanty.

(c) A realistic selection of individual items should be included, even if it is not possible to achieve realistic weighting at this level of detail. (Here questioning of the designer as to his intentions or dreams should be combined with a strong mixture of imagination—say 1:4 by volume.)

(d) The preliminaries bill should be complete in every particular: this is not an area for approximation.

(e) The contingency sum and other provisional sums should allow for the worst that is foreseen and perhaps a little more, depending on the fullness of the quantities. A lot will depend on the realism of the client in appreciating the limits of accuracy of the tender figure.

These points are suggested as the minimum before approximate quantities can be used with fairness to all concerned. In most cases far more should be achieved.

ADVANTAGES OF THE BILLS

The principle of approximate quantities is that of firm quantities in that they still envisage a separation of designer and constructor. They do, however, permit the overlapping of design and construction and this, with the saving in time before tendering, is their main appeal. They also share the general advantages of firm quantities, although to a lesser degree, by giving at least weighted quantities and rates that will usually apply in the final account.

A decision to use approximate quantities does allow some other advantages. The greater expense of firm quantities is avoided—an expense that would be wasted if remeasurement came about in a firm quantities contract. Broadly speaking firm quantities and reasonable variations will attract about the same fees as approximate quantities and remeasurement. The expensive elements in the respective cases are the firm quantities and the remeasurement, and a client who ends up paying for these in combination may have cause for complaint.

It is also possible, with approximate quantities, to include rather more items not actually shown on the drawings than can reasonably be done by using provisional items in firm bills. This can be done by sub-dividing the known items and their quantity, so avoiding inflating the tender amount in the process. This greater range of established rates may be of considerable assistance when the final account stage is reached. There are two facets to this; one is to divide items up into varying sizes and thus help *pro rata* pricing. The other is to introduce items covering fresh types of materials and this will help more particularly when agreeing 'star' rates. This latter facet needs care; if it is overplayed, and an excessive range of materials is introduced, it may mislead the tenderers into assuming that the work is more diversified and piecemeal than is actually the case.

In some cases in which tenderers are asked to put forward alternative design solutions in their tenders there may be a value in having common approximate quantities priced by all tenderers, as a basis with which to assess the schemes proposed and to evaluate work executed. This value will always be somewhat limited, but will be virtually non-existent where the design solutions are not closely similar in the materials and forms that they employ. Also, the original quantities will need to bear a fairly definite relationship to the basic scheme which the tenderers have modified, otherwise these quantities will be of little use as a point of reference. However, used judiciously, the method can be of advantage.

DISADVANTAGES OF THE BILLS
Approximate quantities have a number of clear disadvantages. When the uncertainty as to the character of the work passes a particular point they lose their use, even though they may help beyond where firm quantities have any value. And while they may help where there is genuine uncertainty or shortage of time, there is the danger that their regular use may simply encourage the design team to put off to tomorrow what could well and better be decided today. The danger is not that they can leave decisions until closer to going to tender but that it appears they need not make final decisions until work has started. This is true for measurement, but may not be so for design and construction.

Nor does the nature of approximate quantities help cost control. While the tender stage may not represent anything more than a transient stage in the evolution of the design, it is the best intermediate stage at which to obtain a firm assessment of the cost of the project. Obviously firm quantities are better here, and the more approximate the approximate quantities become, the less use they become. What will count in this case will be the quality of the cost plan or other estimate for the scheme. The level of pricing in the approximate quantities tender may still be useful in reviewing the level of the estimate pricing. Even firm quantities may be of little use at tender stage if it is known that extensive change is proposed and a thoughtful estimate may be nearer the mark.

If the use of bills of quantities for fresh purposes is developed as considered in the following chapter, the approximate variety is of comparatively little use. If the total of the bills is reliable only within limits, how much more uncertain are the individual parts within the bills. This is hardly a criticism of approximate bills themselves, but rather of the circumstances of indecision which they aim to meet. If the job is not designed, the bills can hardly be expected to help greatly in its construction planning, for instance. Even the division of such bills into the normal cost planning elements is seldom likely to be worth while, since the 'swings and roundabouts' of approximation will usually have ignored the boundaries of the elements.

Once approximate quantities pass the tender stage, the actual quantities will usually be of little significance and the rates alone will be of continuing value and will be used as a schedule of rates, in much the same way as a schedule prepared as such and next described. The various standard contract forms have wording that proceeds on this basis.

Schedules of rates

These schedules consist of a list of measured items with units of measurement stated against each, but with no quantities given. The prices may be inserted during tendering, or may already be there, according to the type of schedule as discussed below. A set of preliminaries and preambles, and a form of tender, will also be needed as a minimum.

Of all such schedules it must be said that they should only be used in conditions of even greater uncertainty or hurry than those provoking approximate quantities. Since there are no quantities, someone has to assess the relative proportions of each trade or work section and of the items within them, as best they may. If there are drawings, the quantity surveyor may give his own apportionment of the major divisions as a basis for estimating. It is usually safer, if not always fairer, to ask tenderers to come to their own conclusions on these matters, and thus reduce the avenues of future dispute. Where the data are still vague it is always difficult to communicate any division of them in a form that is neither ambiguous nor misleading. This is not very satisfactory and, as a result, schedules of rates in such circumstances tend to produce much higher unit prices and therefore dearer buildings than do bills of quantities.

Schedules have their uses for 'term contracts', that is contracts entered into to carry out work of maintenance or a series of minor schemes during a fixed period of time, perhaps over two or three years. While exactly what is to be done may be uncertain, the general character of the work will usually be fairly easy to define and tenderers can often assess their level of pricing tolerably well; the detailed work will then be ordered progressively throughout the contract period. With work of this continuing type it will be necessary to provide for the calculation of preliminaries on some sliding scale, to take account of the value of work and also its incidence throughout the contract period. If the work ordered consists of a number of individual jobs of quite different sizes it may also be equitable to have a further sliding scale to adjust the prices for the measured items from job to job. Even for term contracts, it can be seen that schedules of rates are rather in the nature of blunt instruments, but it is often difficult to see an alternative other than prime cost. There are several forms of schedule and the three most common are discussed below.

STANDARD PRE-PRICED SCHEDULES OF RATES

A schedule of this type consists of every measured item that is likely to be encountered in the type of work for which it is designed. Thus even a standard schedule for painting will contain a good number of materials and applications, with all the resultant permutations. A schedule designed to cover all types of work would be a sizeable document indeed; the most widely known of such schedules is that produced by the Department of the Environment. Each item in such a schedule has a price already printed in before tendering begins.

In inviting tenders, it is necessary to issue a qualifying document stating which sections of the schedule are to be ignored by tenderers. These will cover work not expected to be in the scope of the works proposed. For the remaining sections, tenderers are asked to quote percentage adjustments to the prices in the schedule. These adjustments will take account of market factors, the character and situation of the work and the tenderers' competitiveness. Tenderers should not be asked to quote one percentage adjustment to the schedule as a whole, and even one percentage adjustment for each work section may be too crude. Sections like 'Concrete Work' and 'Finishings' will contain diverse elements for which prices will fluctuate up and down in detail between tenderers. The basis of assessing preliminaries will depend on the nature of the project, but must also be given.

The printed prices in such a schedule become out of date with the passage of time, but the percentages quoted can take care of this until the relativities between individual prices become too distorted. The advantages of pre-pricing is that firms tendering frequently on the basis of the same schedule become familiar with the level and detail of its pricing and can, if they wish, fairly easily assess their percentages for the more common sections on past experience. The first time that they tender they will need to price at least the items of major value in the normal way and then compare them with the schedule prices to arrive at percentages. The more prudent estimators will carry out this basic assessment periodically to check their validity.

Just as estimators become familiar with the level of schedule prices, so does the quantity surveyor. This is of value in assessing the competitiveness of tenders.

SPECIALLY PREPARED SCHEDULES OF PRICES

The great advantage of a specially prepared schedule over a standard schedule is that only the items relevant to the job in question will be written into the schedule and, so far as possible, all special and unusual items will be included. Thus the estimators will need to take account of the true job content only. By comparison, most sections of a standard schedule will still contain unwanted items, or those embodying a partly incorrect specification. Obviously the use of specially prepared schedules assumes a job of fairly clear content and usually it will be more suitable to use approximate quantities in such cases.

These schedules fall into two categories, those which are pre-priced and those which are not; the first category will compare most closely with the standard schedule in its operation. In that the schedule will be in use only on the occasion, or occasions, for which it is specially designed, estimators will not have enough experience of it to become familiar with the level or structure of the prices and will not, therefore, be able to shorten their estimating time. Again, percentage adjustments to the prices form the substance of the tenders.

The virtue of issuing an unpriced schedule is that each estimator may price it directly without having to compare his prices with those given to him. He may make what differentials he likes between prices within a section, rather than be bound to take those assumed by the schedule and quote an overall percentage on them. The great drawback of an initially unpriced schedule lies in the difficulty of comparing the tenders received; every price will need individual consideration and detailed quantities will be needed to form a reliable opinion of the order of tenders. The avoidance or impossibility of producing these quantities is the main reason for using a schedule. With a pre-priced schedule the comparison of tenders follows the much more simple procedure set out later in Figure 8.

ADAPTATION OF PREVIOUS BILLS OF QUANTITIES

Where a project follows another closely in character and time, and negotiation with the existing contractor is proposed, the existing bills readily give the body of a schedule if amended in the light of experience. The preliminaries and measured prices in the bills may be amended as needed, either by percentage adjustments or by individual alterations to suit different conditions—or perhaps by a blend of both methods. Both the contractor and the quantity surveyor should find their acquaintance with each other, and with the items and prices, a great help. The quantities in the existing bills will be ignored in the new contract, although they may have been of some value in the negotiation stage; thus the result will be a special case of a specially prepared schedule.

If competition is desired in such a case, a schedule may still be based on the bills but any direct use of the first contractor's prices will obviously not be in order.

Prime cost

The arrangement here referred to as 'prime cost' is also known as 'cost plus', 'cost reimbursement' and 'all daywork' and any fine distinction between these various terms is not worth considering. This arrangement fits into the pattern of this chapter of giving the contract arrangements in order of decreasing precision in telling the client what he is committed to when the tender is received. In the case of this arrangement the uncertainty is at its greatest and concerns not only what is to be built but also what the contractor will spend on building it. The contractor is paid, in the main, what he spends—that is, the 'prime cost'—and only some overheads and profit are usually dealt with on a formula basis.

Since the arrangement carries the penalty of financial uncertainty it is usually employed only when other uncertainties are at their greatest. Thus, it is a useful system when the work is so urgent that it is completely out of the question to allow the time required for any form of estimating related to quantities and for the preparatory documentation. Repairs to dangerous structures are a case in point. It is also the inevitable system when lack of definition of the work precludes the use of estimating in advance. While measurement after the event can be applied to much work, not all of it can be priced on a measured basis. Some alteration work, and some work in inaccessible positions, cannot fairly be estimated in advance even by unit rates in a schedule; nor can work in existing premises where the client intends to dictate the sequence, and the working hours and conditions, from day to day. Some industrial development work, and many maintenance contracts, also fall into this category. Even to price work as measured after execution may mean no more than taking the prime cost and spreading it over the items measured and this is pointless.

There are also clients who think that their best interests, in capital cost and construction time, for more normal projects are served by employing one of a few contractors on a regular basis to carry out their work and to do so by some form of the 'target cost' variant of prime cost arrangements, as discussed below. Mutual understanding, and the benefits of continuity, are felt to outweigh the advantages of measurement as the direct basis of payment. It can also be used in a management contract of the type referred to in Chapter 4: the various firms may be paid on a prime cost basis, while the management contractor's effectiveness is tested by the overall target.

This philosophy seems to emphasise one major fact about prime cost; the 'right' contractor must be employed—that is, one who will seek to work efficiently and thus keep down the sum payable by the client. Thus, while competition over terms is possible in the selection of the contractor the more important element of competition may well be the contractor's reputation for carrying out work economically. A straight negotiation may be the best policy.

There are many variations on the prime cost theme; three broad classes cover most of them and are given below, followed by some points relevant to them all.

COST PLUS PERCENTAGE FEE

Here the contractor is paid for his direct expenditure on labour, materials and plant with a percentage addition on the total of each category to cover his overheads and profit. Where the anticipated cost of the work cannot be assessed this is the only method available; the more that is done the more is paid, both in 'cost' and in 'plus'. An inefficient or unscrupulous contractor who spends longer over the work than he need will therefore secure more profit, since this is strictly related to the turnover achieved. It remains true, however, that many of the circumstances that call for a prime cost arrangement can really be dealt with only on this basis.

COST PLUS FIXED FEE

In this case the contractor still receives full payment for the same items of direct expenditure as before but his overheads and profits are paid as a lump sum or fixed fee, which is related to the anticipated scope of the work and not to the amount of expenditure incurred. Two considerations stem from this arrangement.

Firstly, the scope of the work must be fairly well defined and it must be possible to estimate the cost sufficiently closely for the fee in its turn to be reasonably assessed. Unless this is so, the fee may be too low and may be absorbed in the overheads, with little or no profit for the contractor, or it may be too high and yield an unduly large profit margin. The JCT Fixed Fee Form of Prime Cost Contract recognises the need for fixing the scope by requiring a separate contract for any amendment to it—presumably with an amendment of the original fee (see *BCPG* Chapter 24). But, as already mentioned, to define the scope of work is not necessarily to render it susceptible to estimating in advance. Thus in cases where the prime cost arrangement is unavoidable, and has not been chosen for any other reason, the fixed fee form has its difficulties at the very point where it is so desirable.

The second consideration is the one that makes the system desirable; once the fee is fixed in the contract, whether it be high, fair or low, its effect is to regulate the contractor's expenditure to some extent. While the contractor is still paid for his direct expenditure in its entirety, his fee becomes a lower percentage of this expenditure as the latter rises. Since his overheads will be a relatively constant percentage, this can only mean a lower profit on the turnover. However, the contractor will not suffer an actual loss on the contract until the fee is so thinly spread as not to cover overheads; this is shown in the appendix to this chapter.

This system may be left with a question over it. If an estimate can be obtained in advance, is that estimate going to be sufficiently accurate to render a prime cost arrangement undesirable and bills of quantities preferable? The arrangement also leaves room for haggling over what is a variation and what is a change requiring a separate contract, and the JCT form is not very helpful here. However, the tensions existing under this system are intensified under the next variation to be discussed.

TARGET COST

Under this arrangement the contractor is reimbursed in the first place on the basis of cost plus, either with a percentage or a fixed fee. It is, however, further provided that if the payment due on this basis differs from an estimate agreed between the parties by more than a certain margin either down or up, then the contractor will receive a bonus or incur a penalty as the case may be. This bonus or penalty may be calculated as an adjustment of the fee or as a proportion of that part of the direct expenditure which has been saved or incurred. There is thus a much stronger spur for the contractor to economise than there is under the fixed fee basis, as the appendix to this chapter shows. But there is also a much greater need for accuracy in the initial estimate and in adjusting it to take care of variations. Detailed quantities, accurate but not necessarily based on the Standard Method of Measurement, become inevitable. There are thus fewer cases necessitating prime cost to which it can be applied fairly, and these will obviously be candidates for a measured contract. It may nevertheless be used (as had been suggested) in such cases as a matter of policy, on the basis of the other advantages that it produces for the client. It is also arguable that in some cases the arrangements may give enough incentive to the contractor, and enough control on the client's side, to make it cheaper than a measured contract.

This system also has the advantage, over the fixed fee only system, that it can be operated with a schedule of rates rather than a lump sum estimate as the regulator. It may then be used for projects of uncertain size, which can be measured and priced while the prime cost is being recorded. Here the prime cost will need to be related to percentages rather than to a fixed fee, for obvious reasons.

SOME GENERAL CONSIDERATIONS

In drafting a prime cost contract, or even in adopting a standard form, there are a number of points that should be considered that affect the contractor's reimbursement. The salient items are discussed here but it should not necessarily be assumed that a standard form will have taken full account of them all — this is not the case with the JCT form, for example. An overall aim will be to define what is in the prime cost and what is in the fee. Since these two elements should cover every cost between them, it is best to proceed on the basis of defining one in detail and then defining the other as including 'everything else'; this tends to avoid gaps between the two. The prime cost is the obvious element to define, since it is the element that is built up piecemeal in the final account and also because it is in the contractor's interest to allocate as many costs as possible to the prime cost.

Consideration of which type of prime cost system to use as a financial formula will be related to the factors already discussed, with a view to securing both cost savings and a fair return to the contractor. Two further sets of provisions that serve the same end should be emphasised in the documents. One is to secure the right for the architect to control the level of labour and plant on the site and the categories of labour and plant employed for particular operations, so that any serious wastage may be curtailed. The other is to lay down adequate paperwork procedures for recording, verifying and evaluating the prime cost. If these two post-contract matters are adequately set out, the post-contract phase should become almost a clerical function from the quantity surveyor's point of view, apart from his estimating duties.

The main elements in the definition of the prime cost are given below, while the practical aspects are discussed in Chapter 18.

(a)   Labour: The more troublesome constituents here are those costs which the contractor may incur without effecting any compensatory saving, such as travelling and subsistence expenses, and those costs which may or may not lead to a compensatory saving, such as overtime and incentive payments. In most respects it is best to be quite definite, although some leeway to negotiate items during progress may sometimes be to everybody's benefit. Beyond these elements the main point needing care is the demarcation of those foremen and other supervisors who come under the heading of 'cost' from those who are 'plus'.

(b)   Materials: The place of delivery of materials needs to be considered, as does the pricing of materials to be processed at the works and transported to the site. Labour at the works may be allowed a different rate of overhead, but should it then carry the site rate also by becoming part of the material cost? Materials from stock should be priced at current prices and the permissible level of discount on materials generally should be clarified.

(c)   Plant: This is particularly awkward since, unlike labour, it may not be appropriate to remove it from the site when it is not in use. Apart therefore from having a suitable schedule of rates for plant, there is the question of having two rates for each item; one for working time and one for standing time. Fuel and maintenance are best included in the hourly or weekly rates, but delivery and collection may be better paid for separately. Alternatively it may be possible to pay for large items of plant at a lump sum charge for the duration of their use on the contract: this has all the advantages and problems that attend the use of any lump sum. It should also be considered whether to allow an additional percentage on hired plant and, if so, how to control hiring.

(d)   Nominated work: The only real point here is to define the extent clearly in the absence of PC sums.

(e)   Sub-let work: While sub-letting should require the architect's approval, as in any other contract, it does here pose the problem of whether it is to be paid for on the same basis and at the same rates as though the contractor had carried it out himself, or whether to use the basis of the sub-quotation, with a percentage addition. The prime cost may be noticeably increased if the later system is adopted, as under the JCT form.

Combined arrangements

It is not impossible to use more than one contract arrangement, including some forms of 'all-in service', in the same contract. Some combinations are more whimsical than others and may do more violence to the contract conditions. There are three main aspects where clear dividing lines must be drawn:

(a)   Responsibility for design, specification, site management, the introduction of variations and the approval of work carried out under each arrangement.

(b)   The physical boundaries of work covered by each arrangement.

(c)   The method of reimbursement under each arrangement, and the procedures needed to effect it.

Since there are so many combinations possible it is not practicable to work out the implications of these points in a limited space. Suffice it to say that if the nature of the work or the conditions of its execution are such that the three aspects quoted above cannot be distinguished for each arrangement with crystal clarity, then confusion and perhaps disaster will follow. It is better to think again and adopt a single arrangement for all the work where overlapping exists, using one suitable for all the parts of the work. If it is considered that this will lose the advantage of the arrangements previously proposed, the only alternative is to investigate whether the work can be split into separate contracts. This is hardly likely to be possible unless physical separation exists.

Appendix
A comparison of three prime cost contract arrangements

The figures below show the effect that might be produced in three contracts, all expected to cost £12,000, according to the basis used. The figure of £12,000 is assumed to be made up as follows:

Estimated prime cost	£10,000
Overheads (15% of £10,000)	1,500
Profit (5% of £10,000)	500
	£12,000

PERCENTAGE BASIS
Here the contractor will receive 20% in total on his prime cost, whatever it may be, and therefore will always recover his expenditure and make the same percentage profit.

FIXED FEE BASIS
Here the contractor will receive his fixed fee of £2,000 and the more that is needed to cover his overhead level of 15% of the prime cost, the less will be left for his profit, until actual loss will occur. The figures below show that he can be quite inefficient before this loss occurs but, with the actual prime cost coming out at £13,000 instead of £10,000, he is about to suffer it. (The client still pays the whole expenditure unless any element is specially disallowed.)

Actual prime cost	£13,000
Overheads (15% of £13,000)	1,950
Profit (balance left of £2,000)	50
	£15,000

TARGET COST BASIS

Here the contractor will receive the prime cost allowable and the fee of £2,000; on the figures given below his profit margin is already reduced as a result. But the deduction of half the amount by which these figures exceed the target means that an actual loss is incurred, since the deduction is more than the profit left. Here the effect of the formula is felt much sooner than with the fixed fee basis.

Actual prime cost	£11,000
Overheads (15% of £11,000)	1,650
Profit (balance left out of £2,000)	350
	13,000
Less 50% of £1,000 excess	500
	£12,500

Since actual prime cost plus overheads come to £12,650 and he receives only £12,500 his final position is that he has lost £150 instead of making his paper profit of £350.

CHAPTER 8

The functions of the bill of quantities

The origin of the bill of quantities as a contract document lies in the desire to avoid a number of contractors' estimators having to take off what should be closely similar quantities in order to apply to them the prices which form the really competitive part of the tenders. Its traditional purpose is thus to act as a uniform and labour-saving basis for inviting competitive offers. Its form and content have largely been determined by the needs of estimators and, not unnaturally, quantity surveyors have tailored the document in a number of respects to suit their own ease of production, and also with an eye to the questions of valuations and variations. These somewhat limited, if laudable, aims have left the bill of quantities—so far as the uninitiated are concerned—with something of the mystery of the sausage, while the Standard Method of Measurement has helped to fill out its meaty content with an increasing volume of breadcrumbs. Alongside this, the bill of quantities as a contract document usually acts as the vehicle for the preliminary conditions and for specification information, which are both of wider interest.

In the last decade or two it has become commonplace to question the purpose of the bill of quantities, along with many other venerable institutions in society. Unlike some institutions, however, the bill of quantities looks as though it may survive and may do so by fulfilling a variety of purposes. The seventh edition of the Standard Method of Measurement, currently projected for the mid 1980s, is likely to give things a push in this direction by related proposals for later introducing tiers of rules for such purposes. By adaptation to a changing environment the bill of quantities may thus survive where other dinosaurs have become extinct. Some adaptations have already sprung out of fresh thoughts on how to meet the original purpose, while others are due to the decline of some methods of producing the bills and the rise of other methods. Some, again, have been provoked by the wider questioning and positive demands that a modified bill can meet.

It is said that the quantity surveyor, in taking off his dimensions, produces a great deal of information that would be very useful to others in the building team. Having produced the information, he then ignores much of the detail that he has in the form of notes and processes the dimensions into a form in which much of the remainder of the information is rendered unintelligible to anyone else other than the estimator. If the surveyor could be persuaded to come out of hiding with his treasures, there would be a bonus all round. Since the original tendering purpose is still important, no change should be made which impedes the fulfilment of this; but any of these fresh demands, let alone all of them, if added to the original purpose could produce a bill of an unwieldy character and difficult for the individual users to employ. The answer is to produce several parallel or successive sets of documents, one for each user.

To do this by manual means is a protracted business, although something like it is going on at present in the various parts of the construction team in order to supply what the bill of quantities does not give. Thus the planner carries out one exercise and the buyer another to obtain the information that he requires, sometimes by a rearrangement of the bill in part but often by making a fresh start from the drawings. If the surveyor is to provide the whole of the basic data needed in a readily usable form, but not to perform the functions of those who will use it, then the computer will come into its own. At present it offers marginal advantages of time or money in some cases, as a competitor with other methods of simple data processing. When it can be used to process information into several forms it is likely to dominate the scene. Even more will this be the case if techniques develop to the point where it is possible to use the computer to go from drawing scanning through data banks to tenders with the minimum of manual intervention. The implications of this are however beyond the present work.

Much of the rest of this chapter is an outline discussion of this theme. As such it is really an explanation of what was hinted at under 'Bills of firm quantities' in the preceding chapter and most of it can only be applied to such bills. For the sake of conciseness, the forms of bills which have been developed are summarised in three groups (with a fourth group that does not quite fit the pattern). Then consideration is given to what logically comes first: that is the purposes which are being served or which it is asked should be served. A bill of quantities related to alternative design solutions has been hinted at in the preceding chapter, but there would not be much point in considering it further at the present time as it has not yet emerged in a workable form.

As a tailpiece to the chapter, the matter of preliminaries is considered. These are also logically candidates for an earlier place, but are treated last as their form is little affected by the other matters raised.

The forms of the bill of quantities
TRADE AND WORK SECTION BILLS

These two forms may be considered together, since for present purposes the important point is that they aggregate the quantity of any one measured item, without distinction as to its position in the work or to its timing in the programme; they are distinct from the other two forms in this respect. The only concessions that are usually made are to separate individual buildings in a contract from one another and from the external works, to present the external works in an elemental form and to keep builders' work for specialists distinct from the rest. Otherwise, work is separated only so far as the measurement rules of the Standard Method of Measurement requires.

A trade bill follows the grouping used before the fifth edition of the Standard Method, while a work section bill follows that of the fifth and sixth editions. They are therefore cast in the traditional mould, as indeed is the Standard Method itself. In the case of small projects these bills can usually perform a number of extra functions if carefully annotated, since the content of any one item is fairly easy to place. Even so, they cannot be used for all purposes and the larger the project, the harder it becomes to use them outside traditional boundaries.

ELEMENTAL BILLS

The demand for cost analysis to assist designers, expressed in terms of contractors' selling prices for units of work, has grown in recent years. Two answers have been evolved which start in the bill of quantities—one is to analyse a trade or work section bill and the other is to produce an elemental bill which throws up its own analysis by virtue of its very form. This latter bill will be divided into elements that correspond to functional parts of the project, usually following the standard classification of elements published by the Royal Institution of Chartered Surveyors.

The resultant bill is somewhat longer than the traditional one in most cases, and thus takes longer to write; this is because some items will be repeated in various elements. There is also a slight complicating of the group method of taking off when used. On the other hand, billing can often be started earlier and be done by a direct process. Post-contractually the bill offers advantages by its broadly locational approach. Contractors are, on balance, rather against it at the tendering stage, owing to the repetition which affects sub-quotation enquiries in particular. It is urged that tenderers should be able to produce more accurate tenders, since an elemental bill offers them an indication of where work is and therefore of when it will be carried out. It is doubtful whether many contractors possess sufficiently sophisticated costing systems to produce the relevant data. Even when they do it often indicates sufficient peculiarities in the previous job to preclude the use of the information on a fresh job in the way envisaged. It is also the case that all work in a given element is not always performed in the same time period of the contract while, more importantly, some work scattered in several elements does occur together for organisational purposes. To this extent the elemental bill, like others, may obscure pricing considerations.

The elemental bill has not, therefore, taken on to a great degree, and analysis of a trade or work section bill is the more common method—relating this analysis to annotations or, in some way, to the cost plan. Some combined systems are in use, such as dividing each work section into sub-sections in an elemental sequence, which are more helpful to tenderers. If the sub-sections are printed with each beginning on a separate sheet it is possible to assemble the bill in alternative arrangements for tendering and cost analysis as desired.

Here is an obvious case where computer codes permit the processing of this type of information more readily from a basic bill, or the production of both types of bill from the original dimensions.

OPERATIONAL BILLS

The criticisms of the view that unit quantities can always be related to standard unit prices was referred to at the beginning of the preceding chapter. It has led to the development, on an experimental scale, of bills incorporating either one or two novel features and to the use of such bills on a small number of projects.

The first novel feature is the division of the bill into a large number of sections, each containing a parcel of work which can be executed continuously by the contractor, without the need to break off to perform some other parcel of work. Each such parcel is termed an operation or activity, because it corresponds to those parts of a critical path network to which the bill is related. Within the parcel, a trade or work section order is then followed. The intention is that tenderers should view an operation as an entity and price it as such, taking account of the labour and plant requirements in relation to the situation and stage in the work.

The second novel feature was part of the original concept of the operational bill but is not always employed. This is to present the details of an operation, not in Standard Method quantities but by giving the labour and plant content as a single descriptive paragraph without quantities and giving the materials in the units in which the contractor purchases them. This takes further the pricing of an operation as an entity—indeed, it precludes any other approach. Although certain specially briefed contractors have expressed some sort of preference for this second feature, the general feeling has been that the industry is not yet ready for it on any scale.

Naturally this type of bill is much longer than those in either of the other groups and the proportionate difference in bulk will usually be greater as the size of the project increases, rather than the reverse, since nearly repetitive operations cannot be grouped. Its virtues for valuations and some types of variation are plain and it may have particular advantages when disturbance of progress occurs. The main difficulty with it is that it pre-judges the contractor's method and order of working in the original presentation of the bill; seldom will two contractors work identically. And it does impose this extra burden of pre-judgement on the design team, for drawings are usually produced to tie in with the operations and so this work also is affected. While a reasonably close opinion may be formed on a small project, the problem increases at the same rate as the operations in the case of a larger job. From the point of view of tendering alone, the whole system seems unduly elaborate.

But behind these experiments lies the search for ways in which the bill of quantities can fulfil a number of purposes. What these purposes are will be discussed shortly.

CIVIL ENGINEERING BILLS

This particular category appears in one way to be out of step with the others in that it embraces a particular type of work rather than a particular bill format. It is however standard practice for civil engineering bills, like the external works sections of building bills, to be divided into sections covering each of the functional units in the scheme, such entities as 'sea wall', 'delay tank' and 'approach road'. Within each section work is given in a work section order similar to that in building work. The Civil Engineering Standard Method of Measurement (1976 Edition) now makes such an arrangement mandatory and requires each section to include provisions for pricing separately 'method related charges' that apply to the work in the section as a whole, rather than to unit quantities of individual items.

The end-product of this scheme is a bill that divides into elements, arranged internally in work sections and able to be priced to take some account of the operation as a piece. Hence this separate section given over to the civil engineering bill.

The purposes of the bill of quantities
TENDERING AND SETTLEMENT OF ACCOUNTS

The obtaining of tenders has been referred to in the opening paragraphs of this chapter but this primary purpose should not be forgotten here. It is especially important where competition is involved. From it follows the use of the bill through the post-tender stages, right up to final settlement. All of this is traditional in concept and is expanded in detail in later sections of this book.

DESIGN INFORMATION

There are two forms of design information that a bill may convey. To the contractor it gives specification information; preambles have traditionally given details of materials and workmanship. Positional information may be dealt with in a traditionally arranged bill by annotation of individual measured items, supplemented if necessary by a few schedules in an appendix. A bill arranged in elemental form should reduce the need for annotation considerably, while still requiring schedules. An operational bill should eliminate annotations and curtail schedules.

To the designer a bill can provide cost analysis information; this may be useful to supplement or supersede the cost plan and also for detailed design comparisons. Its main purpose in this respect will be to provide data for future cost plans and the like. For each of these purposes it is positional identity combined with price that is needed for each item and the pattern of usefulness of each form of bill is readily discerned. How far this use of bills can fruitfully be developed will depend on the realism of the information extracted from them, but the stage of development reached to date is such that the principle can now be said to be traditional.

CONSTRUCTION MANAGEMENT INFORMATION

At present, once a contractor has been awarded a contract, his staff have to go back to source to obtain and then process what the quantity surveyor has already worked upon. However, there is some new thinking going on in the industry as to whether the information presented in, or lying behind the bill could be used in additional ways. It remains to be seen how far the seventh edition of the Standard Method of Measurement will give a lead here, when it has been developed.

Construction planning is suggested as one avenue here. Many contractors prepare a fairly detailed programme, even based on some form of network, to assess preliminaries and labour graphs in the tendering stage. For management purposes construction programming is a highly developed technique but it still uses fairly broad quantity data; the operational bill is the tailor-made approach here. The difficulty is for the designer or the quantity surveyor to foresee how the contractor will operate, as has been mentioned already, but a workable method of adapting the initial bill once the contractor was selected could be a help.

The use of the bill for ordering purposes would be a logical step, if it followed on adaptation for programme use, since real assistance in ordering would need to be related to the phasing of delivery requirements. But to provide useful ordering information about pre-formed or pre-cut materials will mean drastic changes in measurement methods, to say nothing of the quality of design information. Quantity surveyors will shudder at the prospect of even trying to provide glass sizes to another party on the basis of drawings provided by a third. For standard units such information can be obtained from manufacturers, but not all buildings are yet composed of standard units. It would be only reasonable, in any case, to leave the problem of mixing allowances and waste where it properly belongs—that is, with the contractor.

Production bonuses are another part of management where quantity is closely involved. However, contractors' policies differ in this: some tie their payments to unit quantities for each man, others to complete sections for a gang. Some will relate their bonus to an analysis of the estimator's pricing, others will proceed from quite distinct reference points. The unions have always resisted any uniform priced measured schedule as a basis. At the present state of bargaining in the industry it is probably best to leave this side of things until a coherent demand is heard.

The preliminaries of the bill of quantities

It is possible to liken preliminaries to hardcore, and that for reasons other than their indigestibility. A contract for construction works of any size at all will need considerable documentation. Much of this will be in the form of technical information, such as drawings and specification data. This part varies from one scheme to another, often quite dramatically, and may be likened to the varying levels of sites. It will also be affected, quite literally, by the character of the site as well as by how the contractor is permitted to operate on it. But in addition to this part of the documentation, there will be a set of contract conditions covering the main legal aspects of the contract. These are comparatively standard from one project to another and are almost invariably one of several standard forms available throughout the industry. As such they may be likened to an unyielding slab of concrete spread over the varying levels of the site. It is the function of the preliminaries to make up levels between the two. In doing so they draw together those elements of information that the contractor needs to price and execute the works and which apply to the works in general and not to some technical aspect in particular. Some will be of the nature of general costs that any site carries to some extent, while others will be obligations or facilities that are peculiar to the job in hand. All are in the nature of an overhead for the total operation.

This twin relationship, to the conditions of contract and to their own technical contents, indicates the main aspects of preliminaries that are relevant to a general consideration. While they are considered here in the setting of the bill of quantities, the remarks are equally applicable to other contract arrangements, although the preliminaries as drafted will not then be able to rely on the Standard Method of Measurement for any definitions that are in that document.

THEIR RELATIONSHIP TO THE CONTRACT CONDITIONS

There is no absolute line of distinction between preliminaries and contract conditions. Thus quite individual matters such as commencement and completion dates are usually in the contract appendix, as are amounts of damages. As between the various standard forms of conditions, there is a variation in the treatment of procedural matters; the ICE form and Form GC/Works/1 both include more of these than do the JCT forms. What is not in the conditions and is relevant must then go into the preliminaries. This is not limited to procedures, and liabilities over nuisance and damage to roads may be subject to differing treatment.

On the other side of the matter, it is not true that every preliminary clause will be specially written for every contract. Perhaps some three quarters of the wording in any one case, if not of complete clauses, will be taken from a common source used for all contracts. Hence the maintenance in most offices of a standard document of some type, taking the form of a substantial bank of clauses to suit related but distinct situations that the particular practice encounters frequently. The National Building Specification contains a section of preliminaries in just such a form with numerous variants, as do several other published documents.

It is the practice to list in the preliminaries the headings of the conditions of contract where these are standard and readily available to tenderers, so that they may be priced. If they are not so available, then they should be issued complete with the other tendering documents, as discussed in Chapter 10. The proposed contents of the contract appendix or its equivalent should also be given since they are most relevant to pricing. It is also possible to include in the preliminaries clauses that are virtually supplementary conditions, either on distinct subjects or as extensions of subjects that are already treated in the conditions. This is particularly the case over procedural matters and is a very useful arrangement. It is also one that needs to be handled with care in view of the status of the conditions. Except in the case of the ICE form, the common conditions state that they are to prevail if there is any conflict between them and the other contract documents (see *BCPG*, Chapters 4 and 27).

When extra clauses given in the preliminaries are clearly supplementary, this provision causes no complication. Where they have an effect that will modify in any way the meaning of a standard condition, they will be rendered ineffectual by such a provision. This will be equally so where a clause in the preliminaries sets out quite categorically to amend a standard condition by deletion or insertion. Such clauses must be regarded as warnings to tenderers of the intended conditions and should be set out as such. When the contract is formalised, the actual body of the conditions must then be amended to incorporate the change to make it effective. It is best to err on the safe side in this respect and make amendments in any case of doubt, rather than to assume that there can be no conflict or modification introduced.

THEIR RELATIONSHIP TO TECHNICAL ASPECTS

Although the preliminaries are present to fulfil a mainly technical purpose, it is easier to draw a boundary between them and the other technical documents than it is to draw the boundary considered above on the legal front. Everything that belongs to the technical description or the physical quantity of some component part of the works in its finished state is a matter to go elsewhere than in the preliminaries. General procedural aspects of testing and approval may be included in the preliminaries but the tests and so forth relevant to any individual part should be ruthlessly allocated to some other place later in the documents. On the basis of this division, the main varieties of items occurring in the preliminaries will be:

(a) *Documentary matters:* As already mentioned, the contract headings are to be given. Lists of drawings issued or available should be given, with any explanatory information that aids the easy assimilation of the documents as a whole.

(b) *Physical matters:* The nature of the works themselves, and any phasing, needs to be summarised and also any factors about the site itself, and how it may be used, over which the contractor might be in doubt.

(c) *Client's requirements:* The client may have laid down particular points that are to be observed during construction that cannot be inferred from the works required as such.

(d) *Contractor's facilities:* There will be various installations that the contractor will need to put in for his own purposes for the duration of the works. He should be able to infer these from the character of that which he is to provide for the client, but he should still be given a classified list of the main groupings, so that he may the more conveniently price them. In the case of bills of quantities these items are specifically extended to include matters of supervision and overheads on labour. It will be necessary to describe any facilities that are being offered to the contractor by the client.

(e) *Nominated and direct contract particulars:* While the individual details will be elsewhere it is necessary to include here any special rules relating to attendance, programme and other working arrangements. The basic rules are covered by the Standard Method of Measurement, sixth edition.

(f) *Procedures:* These may include design procedures over drawings and the like, and site procedures over records and so forth, while on a site within occupied premises there may be particular client procedures over such matters as entry and safety.

Within these varieties there may be a substantial number of individual items to be handled. Each should be so worded as to be complete in itself if it is desired to price it separately, or else should make it plain as to what is dealt with elsewhere. Not all items will permit of such unit pricing, but the pattern of wording should be consistently arranged on this basis. Even where the possibility of separate pricing exists this may be an option which the contractor does not wish to take and he may still include the costs involved in his measured prices.

It can be seen that the types of items outlined above are not affected by any present form of the measured bills. If, however, the bill is to be used in new ways, as the result of future developments, there will be a greater chance of change. This will probably not entail a radical reshaping of what there is at present, but rather the inclusion of a much wider range of items covering the procedural aspects that will assume greater importance in such a situation.

CHAPTER 9

The all-in service

The 'all-in' service is, in its simplest form, a drawing and specification contract in which the contractor is responsible for the design as well as construction—indeed he may even find and obtain the site as well, and clear all the formalities. The reason for the name is quite obvious therefore, as are the other common titles 'package deal' and 'design and build contract'. Another slightly more mysterious label is 'turnkey contract' where the reason appears to be that the client need take only a limited interest until he has the keys and enters. Specialist sub-contractors have been 'giving a scheme and a price' for many years, but the use of this system for the main contract is more fundamental. The subject is given a chapter to itself so that the questions of tendering procedure and contract arrangement may be taken together.

The keynote of this service is that there is no really set pattern. It breaks the traditional structure of designer and builder and permits various alternatives (it can for instance be allied with the management contract approach referred to in Chapter 4, although this does complicate the financial arrangements more than a little). The present survey therefore must be in fairly general terms.

The appointment of the contractor

A client and a contractor may well 'go it alone' in an all-in service. The introduction may have been by recommendation, direct approach or even through an advertisement in a magazine. From then on the contractor acts a as designer and builder until the final settlement. In this case there are no professional advisers and so it need be considered no further.

A client may, however, engage an architect or quantity surveyor or both—and possibly consultants as well—to represent him and watch his interests. If this is done, such representatives should be appointed from the inception of the project, before the contractor is briefed. Moreover, they should brief the contractor themselves, so that they know just who has said what to whom and why; to be brought in later is to be in the position of unravelling somebody's tangled ball of wool. Their initial function will be to clarify the client's ideas and see that the contractor is given a clear and a fair picture of what he is to do. This can well prevent much abortive work, on paper and perhaps on site. If these people are in private professional offices they may need to 'play it by ear'. If they work directly within the client organisation they can usually establish a working arrangement much more quickly.

The contractor's initial brief may be in the form of a performance specification showing what accommodation the building is to provide and what it is to do, much in the form of the brief that an architect himself would expect from his client. Where, however, professional advisers are retained it is likely that they will produce more information than this for the contractor. This will often be in the form of schematic drawings giving an outline planning arrangement. It should be made quite clear to what degree the contractor can put forward proposals to modify such things as room sizes and relationships, storey heights and other major aspects of accommodation. These planning considerations may be affected by the constructional solution put forward, the elevational treatment, or the major services involved. Again, the margin of flexibility allowed to the contractor's designers here should be clear. In principle, only those things that must be fixed should be, so that the maximum opportunities are provided for the contractor's own interpretations. This is particularly important where a contractor is offering a system building or proposes to use industrialised components.

From the aspect of quality control it is helpful to provide a specification, if the contractor is not likely to be sufficiently precise himself. If the contractor's proposed materials and construction are not known, however, it can be a formidable task to give all the alternatives that may apply. Otherwise they can be provided progressively on request as the design develops.

The appointment of the contractor may be the same thing as signing the contract, although there may be an earlier appointment for design work. This will only be in special circumstances, usually it will be a case of 'no contract, no fee'. In many cases the appointment will be reached in one stage by a straightforward offer, either evolved by the contractor alone or crystallising out of a negotiation. Where competition is desired it is fairer to the tenderers not to impose the burden of full design and estimating on them all. The waste of effort in such a procedure is out of all proportion to what is reasonable. It is better to have a two-stage procedure, in the first stage of which the tenderers produce partially developed schemes with an indication of price level and a basis for its further calculation. They do so knowing that there will be some negotiation with several of them in parallel, to determine who is giving the best scheme; obviously they will be in competition over more than price. From this negotiation emerges one contractor, who then proceeds alone to finality in the second stage. This is similar to the pattern described in Chapter 6, 'Negotiation'.

The contract arrangement

The main issue here is the financial risk to be carried by the contractor, including that on account of unknown factors. In the form most obviously attractive to clients the contract will be based on a simple lump sum price with all such risks as ground conditions lying clearly with the contractor. This obviously relies on good information about the strata and artificial obstructions. It is, however, frequently done. If the nerve is not strong enough, a schedule of rates for foundation variations may be incorporated.

Where the package is based on a building system the same arrangement may apply and quite large schemes are let on this basis. But it may be that while the system content is expressed as a package price, the foundations and the incidental pieces of traditional work filling in between the parts of the system are made provisional and are measured on the basis of a schedule. Alternatively these elements may be agreed during the second stage of a two-stage appointment.

There are projects in which it is desirable to have a traditional architect design arrangement for part of the works and an all-in service for the rest. The parts must obviously be clearly distinguishable to keep design responsibility and finance separate, and preferably should be physically separate entities, such as individual buildings. It may be suitable to consider segregation of elements of a building, such as sub-structure in uncertain ground conditions for architect design, or a specialist frame or cladding for contractor design. It is however better for the complete responsibility to be allocated one way or the other, in any case in which a small part might be retained for the alternative treatment. It is also possible to include provisional sums for items like joinery fittings that are localised within the scheme, or even for more diffuse things like services. The latter may lead to difficulties, however, either because of uncertainty over builder's work at the main design stage or because it leaves a split of responsibility over the efficiency of the completed building—if, for example, it is not warm enough. These difficulties can be avoided with certainty only if the design for such services is completed alongside that of the package work and if the designers can jointly commit themselves to accept responsibility.

The essence of the all-in service is that design is complete before the price is agreed. There is however, underlying the fixity of the contract price, what may be termed 'the contractor's dilemma'. He stands to secure the contract on the basis of whether he offers several major advantages, among which design, price and speed are usually salient. While there is tension between all of them, that between design and price is the main one before the contract is signed. The contractor faces competition, actual or possible, over both and has to pitch the level of both at what he hopes is the optimum. There is a similar situation commonly within many industries and in speculative building, but not otherwise in contract building. Here usually the architect has to worry about design efficiency, but he does not meet the construction cost, although he always has the question of professional responsibility and negligence to keep in mind. The dilemma also gives a different slant to the quantity surveyor's activity when he is assessing the tender, as is discussed later under the heading of 'Disadvantages'.

Another strand of the contractor's dilemma is whether he can and should postpone aspects of the detailed design until he has actually been awarded the contract. These questions come out under the headings both of 'Advantages' and 'Disadvantages' hereafter. All being equal, both design and price should be fixed at the contract stage. Otherwise at least one must be fixed, preferably price, and the other must be quite closely constrained. If any element, such as foundations, is subject to measurement and payment as executed, there is also a need for some mechanism in the contract to ensure that there is not an over-provision physically by the contractor, while he still retains design responsibility so far as possible.

The JCT standard form
Several standard forms of 'design and build' contract exist. They do not assume any particular tendering procedure, although certain groups of documents are required in each instance. That produced by the JCT is the Standard Form of Building Contract With Contractor's Design (see *BCPG* Chapter 22). It names the client and the contractor (the former as 'the Employer') and allows for the appointment by the client of an agent, or possibly agents, solely to look after his interests. It does not specify any consultants acting for the client. This gives the position shown in Figure 6(iii) in Chapter 4.

The JCT form accommodates the general philosophy of the all-in service within a framework so far as possible similar to that of the other JCT forms, but introduces three particular different elements. The first is the employer's requirements: the nature of these is not prescribed, but they must include all essential briefing information and other matters over which it is desired to limit or eliminate the contractor's freedom of choice. The second is the contractor's proposals: again the nature is not prescribed, but whatever is not determined by the employer's requirements must be taken up and defined in the contractor's proposals. As between these two elements the contract is quite unrestrictive, and so there may be great extremes over the distribution of responsibility for producing information between the parties also. There is also provision for the contractor to complete aspects of design after the contract is signed, but without financial adjustment unless it falls within the scope of a provisional sum. There is no provision for nomination under the form and it should not be attempted.

The third element is the contract sum analysis. Once more the nature of this is not prescribed by the contract, but it is to be in the form given in the employer's requirements. Reasonably it may vary from a number of lump sums to some sort of priced quantities, according to the complexity of the project and the intended use of the analysis for purposes of variations, interim payments and perhaps formula fluctuations. The assumption is that, provisional sums apart, the contract sum is firm and that the contractor carries all risks of design development and of ground conditions and other such hazards. Balanced against this, the right of the client to require variations (termed 'changes') is limited to cases in which the contractor does not object that they are unreasonable. Problems over an analysis of this type are mentioned under 'Disadvantages' later, but it is clearly something that the quantity surveyor will wish to see established in the best form possible.

A hybrid arrangement for design of part of the works by the architect and part by the contractor is possible by using another document, the JCT Contractor's Designed Portion Supplement (see *BCPG* Chapter 23). This is used in conjunction with the JCT form with quantities and introduces amendments to bring in an all-in service for the portion concerned, while leaving the rest of the scheme under the usual arrangements. This may give a useful measure of flexibility in circumstances already discussed.

Advantages of the all-in service
Several of the advantages claimed for the all-in service flow from the fact that design and construction are undertaken by one organisation; the traditional division of the industry in this respect has been referred to in Chapter 5. Where there is one organisation, past experience should be blended into current designs, but much will depend on the integration within the organisation and the feed-back techniques. It is of course true that independent designers also learn by experience, from a different angle. Some contractors, in offering an all-in service, rely on commissioning private architects and this would seem to nullify the apparent advantages.

But since there is competition over the design as well as the construction, an integrated all-in service may achieve a result which is cheaper per unit of accommodation provided. This is likely to happen only if the contractor is good competitively at both design and construction and not just at one of them. Because the contractor is paid only one sum of money for his two functions there should be every incentive for him to optimise the proportion of his expenditure committed to each. Some projects warrant a higher design element than is represented by professional fees to give a construction economy, while others can be done efficiently with less. The result should therefore be a lower contract price.

The arrangement is most useful for buildings with a fairly simple form or function; something like a farm storage building lends itself well to it, since the amount of briefing and the design element are both quite low. Where a system building is called for, some form of package must be employed as the central part of the contract. Again, the farm building may come into this category, but so too will many housing schemes.

An all-in service is often quite quick, from initial brief to handing-over of the finished building. This may well be because the client emphasises speed and chooses a package because he mistakenly thinks a traditional approach is bound to be slower. And firms giving an all-in service tend to put this aspect first, if asked to, rather than to be squeamish over design considerations and procedure.

Finally, whether the all-in service is cheaper or not in a particular case it is, of course, really 'all-in'. There is no string of meticulously calculated and lengthily presented fees in its accounts to disturb the equanimity of the client. It looks tidy and does not highlight where the money goes.

Disadvantages of the all-in service
Very often the all-in service removes the element of competition entirely, although the client's advisers should be able to assess whether value for money is being offered. Even when there is first-stage competition this will leave a lot to the second and non-competitive stage. It may therefore be that while the design is economical by reason of its standard of design it could still be built more cheaply, for each unit quantity of construction, by someone appointed competitively under the traditional processes. The real question, however, is whether it would have been designed so economically as well. A simplified example may make the point:

(a) Architect designed scheme, priced by contractor
       100 unit quantities of construction
                          at £5.00 per unit = £500.00 contract sum

(b) All-in service scheme, to meet same brief
       90 unit quantities of construction
                          at £5.25 per unit = £472.50 contract sum

It often happens that the building is not fully detailed at the stage when the contract is signed. There is a natural reluctance on the contractor's part to do work which will be wasted unless the scheme goes ahead. Also it is one way of saving time, which after all is one of the package's strong selling points. The details concerned may well be made 'subject to the architect's approval' where an independent architect is retained. But the end result may be unsatisfactory for one or both of two reasons. First, there is always the possibility that the client finds that he is not getting what he imagined he might get—not that this state of affairs is unknown in more traditional arrangements. Secondly, it may also happen that the contractor not only did not draw his details but did not consider what he was going to do and, in consequence, he put a cover price in his tender; he will obviously hope to design within that figure but the result may hit either party.

A bigger snag may be that, with a competitive design, the design will tend to be developed primarily with capital cost in mind. In this the design may revolve around the contractor's own favourite methods of construction. As a result the building may suffer in its functional arrangement or its aesthetic quality. It may not be cheap to operate, or it may have high maintenance costs and these considerations will not have the same force in a contractor-designed building. However, provided that the client and his advisers brief a single contractor clearly, and keep in touch with him, things should not go too far astray, while in a competitive selection between several contractors more than capital cost will be under consideration. But it is not always easy to correct the bias in a partly developed scheme, even when this is attempted soon enough.

Because the all-in service is based upon a lump sum price without quantities, it is more difficult to secure a basis for variations. This is partly because the nature of the contractor's design solution may not be determined when the attempt is made to prescribe the form of the basis, so that the materials and so forth are uncertain or even the layout and elements of the project are in doubt. The JCT practice note recommends a breakdown into quite a number of lump sums to cover the most likely elements, and this could include some form of quantities in suitable cases. Otherwise any of the methods discussed under 'Drawings and specification' in Chapter 7 may be used, but they become less effective as the size of the project increases. Thus the client is at risk and may have to pay heavily for changing his mind—but there are many, and not all of them contractors, who would consider this a positive advantage.

The place of the quantity surveyor in the all-in service

What has been said in this chapter has not necessarily been the direct concern of the quantity surveyor, although he should weigh these considerations when giving advice or dealing with the all-in contractor. Which of his skills he will need, if he is introduced into the operation at all, will depend on the precise pattern adopted. The most likely duties will be among the following:

(a) Feasibility studies and budget establishment.
(b) Outline cost planning in preparation for briefing the contractor or con-
    tractors; detailed cost planning and checking with the selected
    contractors.

(c) Drafting tender documents and contract details, or checking contract proposals from contractors.

(d) Whole life cost appraisal of outline schemes from contractors in competition, with alternative estimates and negotiations required to amend any of the schemes.

(e) Negotiations with the selected contractor and recommendations on his tender and its analysis.

(f) Normal post-contract activities, on whatever basis the contract provides.

The earlier the quantity surveyor comes on the scene, the more usefully can his contribution influence the project development. The philosophy of the all-in service tends to emphasise good faith and to require a certain flexibility of outlook and dealing if it is to be successful. Of these, flexibility is not always the quantity surveyor's predominant characteristic, due to the terms of reference within which he normally operates. Hence the occasional package may be salutary.

CHAPTER 10

Some principles relevant to pre-tender activity

There are many matters that do not fit neatly into the pigeon-holes of the preceding chapters in this part of the book. The present chapter is designed as a sort of ragbag to collect up the more important. They mainly concern tendering procedures and contract arrangements discussed previously but are considered here directly in relation to competitive tendering and bills of quantities, so as to avoid repetition. The differences needed to meet other cases are fairly obvious and easily deduced.

Parity of tendering

It is axiomatic to competitive tendering that all firms competing shall receive the same information from the client's advisers and be given the same time and facilities for preparing properly considered tenders. Apart from the initial documents sent out to tenderers, it often happens that some tenderers raise queries during tendering and receive additional or modified information by way of answers. This information should be put into writing and circulated to all tenderers, to maintain the parity desired. Again, a tenderer sometimes wishes to put forward an alternative material or process and there should, in general, be no objection to this so long as the tenderer puts forward his tender on the original basis and includes a covering letter giving details of his alternative and its financial effect.

While the principle of obtaining tenders strictly in accordance with the terms of the invitation should be adhered to, there is much to be said for positively inviting tenderers to put forward their own suggestions for amendments within clearly defined limits. These may be matters of specification but may also embrace changes in the contract period or sequence of working that offer financial or other advantages. The unmodified tenders will then serve to establish the overall order of tendering, with the advantages of any individual modifications offered. If a tenderer has not put forward a modification in a way suggested, then presumably he has none to offer and justice will have been done.

The position where a tenderer has taken the initiative in putting forward an amendment is a little more delicate. A radical suggestion may be of such consequence that it is tempting to see what other tenderers make of it and yet the originator stands possibly to lose the benefit of his originality. This sort of case can only be dealt with on its merits but certainly, if the suggestion is to be circulated, the originator should be told. (If the suggestion should involve a proprietary process or system there will be legal considerations to be taken into account; the same will apply if the suggestion is in the form of a design which may be subject to copyright.)

There are two ways of dealing with the contract period when inviting tenders. One is to stipulate a definite period and the other is to leave it to tenderers to state their own desired period. While this second method does not destroy parity of tendering, it does leave the tenderers in some doubt as to whether price or time is the main factor that will be considered when tenders are received. If it is used some guide lines should be given but in most cases the naming of a definite contract period when inviting tenders is to be preferred.

Documents required for tendering

The question of which documents to send out to tender, and how many of each, may be decided for the quantity surveyor if he is dealing with an official client; the latter may even do the issuing as well. If the surveyor happens to be within the client's organisation he may of course be in on the decision making. Whoever calls the tune, there are several points worth noting.

CONTRACT CONDITIONS

Contract conditions nowadays are almost invariably one or other of the standard documents, which are readily obtainable by any tenderer. Following the line of the Standard Method of Measurement, therefore, the quantity surveyor need not include these conditions in the issued documents. Any proposed amendments to the standard wording, however, must be set out in full in the preliminaries, along with a list of clause headings, as must also be the data to be filled into any appendices or schedules. The use of entirely non-standard forms of contract should be resisted except in the most extraordinary circumstances; if they must be used they must be sent in full to all tenderers.

THE SPECIFICATION

The specification is not recognised as a separate contract document by the various JCT 'with quantities' conditions and its meat is usually included in the preambles to the bill. Any positional information is then issued when the work starts, in a non-contract document. A very lengthy and involved specification, particularly of the type produced by a consultant, may best be dealt with physically as a document bound separately from the bills and then incorporated into the bills by cross-reference. Under the GC/Works/1 conditions the specification is a distinct contract document and it will be necessary to cross-reference it and the bills throughout. Such cross-reference is not always as simple as it sounds, since the contents of specifications are often arranged with other ends in mind. Any cross-referencing may therefore need to be selective in its approach and perhaps to re-state some things that have already been said.

If the ICE conditions are used then, by virtue of these and the Civil Engineering Standard Method of Measurement, the specification and drawing will be relied on much more heavily in the bills, which will be correspondingly shorter.

DRAWINGS
It is good policy to issue some drawings to tenderers with the other documents. These drawings will mainly be the general arrangements drawings and any other drawings needed for tenderers to assess the constructional approach that the project will demand. The larger or more intricate the building, the more of these drawings there will be. In addition there will be the site layout drawing and any drawings governing phasing and other special conditions. In short, there should be all drawings affecting price in ways that cannot be adequately covered in the wording of the bills of quantities. All these drawings will rank as contract drawings later. Drawings that simply show detail need not be issued but should be on display at the various professional offices for tenderers to inspect if they wish. Whatever happens, the tenderers should of course be left in no doubt that measured items in the bills are to be priced as described and quantified in the bills, even when the items or the bills as a whole are approximate.

An alternative philosophy to the foregoing is not to send out any drawings at all and to leave it to tenderers to inspect them all at the professional offices. This saves a little cost, obviously, but it is a false economy; estimating time is wasted in otherwise unnecessary visits, and a hurried flutter through a pile of unanalysed drawings may lead to either costs or economies in construction being overlooked.

Where comparatively specialised work needing fabrication at works has been measured in the bills there will be a special case for issuing detail drawings to tenderers. Such items as special metal windows, pressed metal work, fittings and plant cannot be priced from bill descriptions and sketches only, be the quantity surveyor never so lyrical. It is better to describe the scope of each item briefly and what is excluded at its boundary and then to 'reference in' the related drawing to the bills. The only problem here is the number of copies that tenderers may require for sub-quotations.

BILLS OF QUANTITIES
To repeat what was said immediately above, the only problem here is the

number of copies. The traditional approach has been to send one to each tenderer and then to ask the tenderer whose tender is initially under consideration to copy his prices into a further copy for examination. The only objection to this is the delay of two or three days that takes place at a vital time, especially if the client is breathing hard.

To avoid this delay it has sometimes become the practice to send each tenderer two copies of the bills and to ask for one to be returned priced at the same time as the form of tender but sealed in a separate envelope. Only that of the tenderer or tenderers under consideration is opened and the rest are returned unopened. On balance this procedure does not seem worth while; apart from the marginal cost to the client of extra printing, the system does involve all tenderers rather than just one in the extra labour of filling in the additional copy—and that is in the hectic closing stages of tendering. Often sub-quotations are not to hand in detail to insert in the copies and sometimes there also lurks the suspicion that the prices of tenderers not under consideration are examined all the same, to gain an extra check on those being considered.

Nowadays extra copies of sections of the bills, for obtaining sub-quotations, can be photo-copied easily enough. It is, however, sometimes the practice to print extra copies of these sections with the main run of the bills, after obtaining from the tenderers their anticipated requirements; it has been recommended in an RIBA practice note that these copies be charged at net cost. Whoever prepares the extracts, it is vital that the associated preambles and any governing conditions are extracted as well.

THE FORM OF TENDER

The form of tender is not part of the JCT or ICE conditions but a letter accepting the offer made on the form will be evidence of a binding contract until the formalities have been completed. In the case of the GC/Works/1 conditions the form of tender is usually combined with the invitation to tender and an abstract of particulars, both of which set out important details of the contract. Here also the contract is formed by an exchange of letters related to the form of tender, which thus becomes a contract document.

The form should state the names of the client, the architect and the project and should refer to the tendering documents in outline.In the case of the JCT forms, where fluctuations apply the date fixed for the receipt of the tenders is important and should be given on the form. The tenderer should sign and date the form, giving his tender amount in figures and words and also giving the contract period if this is not already included. A ready addressed envelope should have been provided for the return of the tender, stating when it is to be received but giving no indication of the sender.

It is a useful plan to bind an additional copy of the form of tender into the back of the bills, for record purposes only.

The use of prime cost sums and provisional sums

PRIME COST SUMS

Prime costs sums, usually abbreviated to PC sums, are a useful way of writing into bills of quantities various sums of money which tenderers will include in their total tenders but which are to be reserved for expenditure on work to be done by specialist firms chosen by the architect later. The initial effect is that the client accepts the tender, which is inclusive of the cost of all work required for the project even though part of it is to be sub-contracted. The Standard Method of Measurement in its sixth edition has the following clauses relative to PC sums:

Clause B9:   Works by nominated sub-contractors — The name of the firm is to be given, and a description of the work, along with items to cover general and other attendance.

Clause B10:   Goods and materials from nominated suppliers — The name of the firm is to be given, if known. A definition of 'fixing' by the contractor is included.

Each of the standard forms of contract has its own rules and procedures governing PC sums and these differ in quite a number of important respects—not least over the question of discounts. These forms and the procedures they incorporate with regard to PC sums are discussed in *BCPG* Chapters 15, 16, 27 and 28.

An essential feature of PC sums is the question of nomination, that is the right reserved to the architect or his equivalent to name the firms which are to carry out the work represented by the PC sums and then to instruct the contractor to enter into a sub-contract with each firm on the basis of terms already obtained by the architect. While the architect—and in many cases the quantity surveyor—will have prepared documents and invited tenders for the work, and will have examined the tenders and selected the most suitable, they then drop out of the picture and it is the contractor alone who is party to the sub-contract and who thus has privity of contract with the firm concerned. The client has no relationship with any sub-contracting firm, unless the employer/sub-contractor agreement or the supplier warranty is used to give a collateral relationship outside the direct line of the contract (see *BCPG* Chapter 17). His advisers will however be responsible for settling with each firm the amount to be paid in the final account if this is different from the original sum. This is usually expedient to save three-tier negotiations.

A variant on the PC sum as such is the PC rate embodied in a measured item. Here the precise specification of the main component or other material content need not be given and the tenderer does not obtain a special quotation. Instead the tenderer is given a price for this part of the item, for inclusion in the measured price for the complete item. This method has no particular advantage for a single component item such as a lock but it is useful for materials like bricks, where the wastage element is a significant factor.

The value of the PC sum or rate arrangement is that it gives a precise control over the process of selection and over the quality of the work. There is a limit to what can be specified in words. Just as a main contractor is best selected from a short list of known firms to secure quality of work, so it is at least as important to select a specialist sub-contractor by a similar means. Merely to specify and measure specialist work in the bills and then leave selection to the contractor may be most unfortunate in its outcome. There are also cases where the architect will need advice from the specialist firm, and perhaps even a complete design, before he can complete the design of the building as a whole. In these cases it may come about that the specialist will be chosen before the main contractor and will be ready and waiting to be nominated. There is also the advantage that the design work can be paid for under the PC sum expenditure without introducing any extraneous accounts; this is quite a permissible approach where the appointment of a consultant is not justified.

The system also has its uses, in what is sometimes a more dubious way, to enable design decisions to be postponed. Once the sum is written into the bills, just what it represents can be decided later. There may be sufficient reasons for this but it may arise just out of lethargy and be resorted to out of habit. Good reasons for such delay in deciding would arise where the relative economics of alternative materials were likely to fluctuate or where the sub-contract work was far ahead in the building programme and more realistic prices could be anticipated nearer to the time.

In recent years the tendency for PC sums to account for over half the total sum in many tenders has led to a questioning of the extent of the system. It is said that contractors have less control over nominated firms than over firms of their own choosing. This is not something inherent in the system but often arises from weakness in applying the factor of privity of contract by all persons concerned. Again it is said that contractors obtain lower quotations than do architects and quantity surveyors, perhaps by 'dutch auction' tactics; this must be weighed against the other controls desired. For many types of finishing, inside and out, and other work that has now become reasonably standard to specify, it would seem more straightforward to dispense with PC sums. But for the really specialised work, and where special considerations predominate, they are of real value.

PROVISIONAL SUMS

Provisional sums may be abbreviated to 'PSs' and this perhaps is how they come to be regarded, acting as a last minute way out of all the unresolved problems. They have a better claim to fame, however. Like PC sums they help to ensure that a tender is inclusive of the full weight of the client's commitment, though beyond this they are simply sums of money awaiting the architect's instructions about their expenditure. They are used where some part of the work which is known to be required is uncertain in character; whereas PC sums may well be for work that is of known character. Furthermore, provisional sums have a partial distinction from PC sums in that they may represent work which will ultimately be carried out by the contractor or by a nominated firm. As such they are to be read as not including any discount or attracting any profit addition.

The sixth edition of the Standard Method of Measurement defines provisional sums in Clause A8.1a and each of the standard forms of contract recognises their use for a number of purposes. All of these references add up to the same general picture of including a sum of money in the contract so that the client and the contractor both have the full story. The contractor can therefore make his level of preliminaries realistic, since these must cover all the work delineated in the bills, no matter how it has been presented.

The use of provisional sums should be as niggardly as possible. They should be employed only for small pockets of work, where the character of the work or the conditions of working are so uncertain that no price can be estimated in advance. When once a provisional sum is included the work represented is removed from competition or pre-tender negotiation; if therefore it is possible to use provisional quantities rather than a sum this should be done. A sum should not be used solely because the extent of the work is uncertain, or because the rates obtained may need to be adjusted in the final account. It is always better to have *some* rates, rather than none as the point of departure.

There are some items, like insurance premiums and fees and charges, where the provisional sum is the only answer if a competitive price cannot be given. Another, and special, case of a provisional sum is the amount included for contingencies.

THE CONTINGENCY SUM

In building, as in many walks of life, the unexpected is always happening. In even the most meticulously planned contract there should be a contingency sum written in to give a shock-absorber against the unforeseen. It is difficult to generalise on its level but for new work it is often between 2½ and 5 per cent of the total value, while for alterations work 10 to 15 per cent may represent the uncertainty better. Where quantities are approximate the upper level of contingency will often be regarded as a minimum. What is actually chosen in such a case will depend on how realistic the quantities are, or even the degree to which they have been deliberately inflated.

No quantity surveyor should write in his own idea of the contingency sum but should discuss the matter with the architect. The client is also entitled to know the level of what has been included, and why; if he is an official client he may even have fixed rules which dictate the level. The principle is that the client is giving a margin to the architect with which to take up inevitable extra expenditure, or the afterthoughts of either of them. The client therefore should know what he is starting with, so that he may keep track of how it is going.

It is often the case that PC sums will be overestimated to give the architect a little bit of extra contingency; if the quantity surveyor wants his share he has to look to the provisional sums and quantities and not to any weighting of the firm quantities. The inclusion of these hidden contingencies is a two-edged sword—they may save minor recriminations when dealing with an extravagant architect or client but if the tender comes in too high because of them they may cause misery at an earlier stage.

Fluctuations or cost variation
Before discussing the merits of allowing the contract price to be adjusted to meet changes in the prices of labour or materials, or asking the contractor to allow for them in his tender, it is worth remarking that a 'firm price' contract is simply the alternative to a 'fluctuating price' contract. The firm price contract may be related to approximate quantities or even a schedule of rates but its individual prices are firm even if the total is approximate or non-existent. The term 'firm price' is sometimes used as synonymous with 'fixed price' which may be distinguished however as meaning a stated total contract sum as against a sum to be ascertained after completion of the works by remeasurement or prime cost.

The method of dealing with fluctuations (termed 'cost variation' in government circles) is regulated by the particular contract conditions. These produce comparable effects, although there is a fair difference in detail between them. In the present section discussion assumes the JCT clauses as far as there is any difference. There are two critically different approaches to calculating fluctuations in the post-contract stage, and it is necessary to outline them here as the approach must be selected before entering into the contract — in fact the contractor must be clear as to which is to apply, if either, before he can prepare his tender.

TRADITIONAL FLUCTUATIONS

The more traditional approach to fluctuations (see *BCPG* Chapter 10) is to calculate the increase or decrease in costs actually incurred by the contractor for labour, materials and energy owing to changes in both market prices and statutory payments, or alternatively in statutory payments alone. These alternatives are often known respectively as 'full' and 'limited' fluctuations and also permit some adjustment to their operation to be built into the contract to narrow their scope. Beyond this possibility, they require little action at pre-tender stage other than a suitable appendix in the bills or attached to the form of tender, for the list of materials and fuels (if any) which are to be subject to fluctuation adjustment.

There is nothing express in the standard provisions to cover fluctuations in costs of plant, overheads and other elements not defined as in the foregoing. However, there is available a percentage addition to be made to the reimbursable items that covers these costs in a blanket way. The alternative method considered next effectively embraces all heads of cost.

This approach is somewhat less involved in its basic concepts and can be used for contracts based on drawings and specification, as its rival cannot be. It also has some flexibility in what it allows to be adjusted and over how the adjustment is performed, as is discussed in Chapter 19. Against these considerations, it does make heavy demands on staff time for both contractor and quantity surveyor because of its reliance on detailed investigation of time sheets and invoices. This tells against it on the larger contract in particular. It also leans on costs incurred, rather than work produced, for its basic information, so that the amount that the client pays is influenced somewhat by the contractor's working efficiency or the markets in which he buys his inputs, as suggested in Chapter 5. Both these aspects are considered in Chapter 19 in relation to the calculations entailed.

FORMULA FLUCTUATIONS

The newer approach to fluctuations (see *BCPG*, Chapter 10) is to calculate an increase or decrease which bears a relationship to the measured prices in the contract, and not to the costs actually incurred by the contractor. The adjustment is therefore based on output and not input. The method is usually known as 'formula fluctuations' for reasons that results from the following main elements that are delineated in the published Standard Form of Building Contract Formula Rules as necessary to its operation:

(a) Bills of quantities are divided into sections or items are annotated so that all measured items are identified within a series of work categories (or in simple cases, broader work groups). The categories are similar to bill work sections but rather smaller, so that each embraces bundles of work with roughly comparable proportions of labour, materials and plant and common types of each as between the various items. This should mean that the fluctuation on any two items within one category over the same period should be comparable and thus permit the same level of adjustment.

(b) Monthly bulletins are issued by the Property Services Agency of the Department of the Environment giving index numbers for each of the work categories. Current measured price information is collected by the Property Services Agency to form the basis of these index numbers and thus the changes in the numbers from month to month indicate the changes in prices being tendered. Such changes are partly due to changes in the cost of individual inputs without limitation to labour and materials alone, but also reflect other factors such as seasonal influences and the state of contractors' order books.

(c) A base month is defined in the contract as that preceding the date of tender, which is also specially defined. The index numbers in the bulletin for the base month are deemed to be those on which the tender is based. It is necessary to be quite clear on the relationship of the date and month, since at a particular time in a calendar month there will be a shift in the base month.

(d) As work proceeds, interim valuations are prepared at monthly intervals and on the same day of each month. Valuations are divided into work categories and the additional value of each work category for the month is extracted at contract rates.

(e) Each of the additional values as isolated at the contract price level is then entered into a formula, the essence of which is:

$$\text{Additional value} \times \frac{\text{current index} - \text{base index}}{\text{base index}}$$
$$= \text{amount of fluctuation in category for month}$$

(f) The fluctuations amounts are aggregated throughout the work and are final amounts, open to adjustment for a limited range of matters only, as discussed in Chapter 19. Any balance thrown up in final settlement that adjusts the interim values in total is treated by an 'averaging' provision.

(g) Values other than for measured work are given particular treatment, either by a further averaging provision or by exclusion from the formula operation and the use of a direct cost basis.

Clause 40 of the JCT form refers to these rules. The system means that the contractor is paid on a basis which may be out of step with the costs that he, in particular, incurs. The result however is that the contractor is paid amounts related to the original price basis which are not influenced by his operating or commercial efficiency, but simply by national trends. The system may be considered somewhat insensitive in that it ignores regional trends. It does however avoid much tedious work in calculating fluctuations by all concerned, and this is the reason for its introduction. On the other hand it does require quite elaborate clerical work of its own. The system is likely to be reviewed in a number of respects, but is equally likely to gain in favour.

The approach is not intended for contracts which do not embody quantities or for those covering complex alteration work, since in each case financial segregation of work categories or groups is not clear. When it can be used with quantities it does depend on careful, regular valuations, since the effect is final. The method can only be used as given and is a 'full' fluctuations approach that cannot be restricted to some elements of fluctuations and not others.

WHEN TO USE FLUCTUATIONS

Whether to allow fluctuations or not is a policy matter, to be decided primarily by the economic climate. The larger the contract, and the longer the contract period, the greater will be the uncertainty imparted into the assessment. Many contractors will press for fluctuations whenever possible, since a risk element is thereby removed, but there are those who may wish to avoid the disproportionate trouble of recovering increases. If contractors are working to a firm price basis there is an incentive for them to resist increases from men and merchants, and thus strengthen any economic stability that there happens to be; from this point of view, firm prices are highly desirable. Since the last war it has not been possible to avoid fluctuations entirely over really long periods, and then only for contracts due to be completed within two years. In recent years the inflationary spiral has made their use inevitable.

If a fluctuating basis is adopted for a contract then the scope of the fluctuations should be kept as tight as possible, on the foregoing principles. With the formula method there is of course no chance of narrowing matters. With the traditional method there are several avenues for restriction. There is the move from full to limited fluctuations for either labour or materials or for both of them. More drastically fluctuations could be restricted, at one level or the other, to only labour or materials. Fluctuations on materials should in any case be confined to as small a list of materials as it reasonable — these being the ones of consequence in quantity in the work or those whose market prices are prone to vary violently. Once labour is made subject to fluctuations there is little further that can be done, since what is allowable is spelt out by the contract conditions, and it is unwise to alter these.

There is another approach to the question of 'to fluctuate or not to fluctuate' which can be used more when the market conditions are equivocal. This is to invite alternative tenders on a fluctuating basis and on a firm basis. If this is done for a measured contract arrangement in which total sums are tendered, as distinct from a schedule, the unit prices should be related to the more favoured possibility and tenderers should be asked initially to give a lump sum adjustment for the other possibility; if this were to be followed up it would then be worked out in detail. This does cause more work for tenderers and also creates a further problem of assessment when tenders are received. Since the order of tenders may be changed, no tenderer can really afford to pitch his firm price too high to ensure that the other is accepted.

A closing thought is that fluctuations may be restricted to those which exceed a margin of say 2½ per cent, or some other margin that is not expected to be exceeded to any extent, in times of relative equilibrium when only the excess would be adjusted for any category like wages, insurances or individual materials. This could be applied under any version of the fluctuations provisions to the whole of the fluctuations or to any selected sub-division of them. It would have the effect of setting up a firm price contract that would become a fluctuating price contract beyond the stated limit. Such an arrangement is to be distinguished from that built into formula fluctuations for public contracts, under which fluctuations are adjusted whatever the change in the index but the gross amount of adjustment is abated by a fixed percentage on grounds of increased productivity.

PART 3

THE EXAMINATION OF TENDER STAGE

CHAPTER 11

Examination of tenders based on bills of firm quantities

When once the bills of quantities are safely off to the tenderers, a measure of calm descends on the quantity surveyor's office—at least so far as that particular project is concerned. The remainder of the printed copies are stored away, after any further distribution to the architect and to the office files. All the scattered papers are collected up and there is little to do but wait to hear which tenderers have received the wrong covering letter or what mistakes have occurred in the reading over.

It is important during the tendering period that parity of tendering is maintained on the lines considered in Chapter 10. The tendering period itself should also be of adequate length. For a project of reasonable size there will be several stages to the tenderer's estimating process, including that of waiting for quotations to be received from suppliers and potential sub-contractors, and four weeks should be allowed as a minimum. On very large projects, or where design work is involved, a longer period should be given.

It sometimes happens that a tenderer requests an extension of the tendering period. If only one tenderer does this it may be better to hold him to the original date, and then double check his tender if it is the lowest! If, however, several tenderers make the same request it is obvious that not enough considered tenders may be received and an extension may have to be conceded. Anything more than a week is hard to justify on grounds of difficulties with quotations or holidays, assuming that the original period was adequate. Where a committee date must be met, or the client is agitating for work to start, it may be necessary to deny any extension or at least to limit it severely.

If an extension is granted it should be notified in writing and tenderers should be asked to confirm receipt, as they should with all other amendments as well as with the original documents. Moving the date for receipt of tenders will also move the nominal date of tender for fluctuations purposes. While this will only be the same number of days shift when the older JCT clause 38 or 39 or an equivalent is used, there could be a month's shift when JCT clause 40 for formula fluctuations or some similar clause used and the tenderers' attention should be drawn to the change from the month stated in the contract appendix information given in the bill preliminaries.

The receipt of the tenders

It is sound practice that a panel of people should open tenders, to avoid any suggestion of irregularity in proceedings. Where the client is a committee the composition of such a panel is clear; in other circumstances the client, the architect and the quantity surveyor make a convenient group. The old practice of having the tenderers present has fallen into virtual disuse. There is an advantage in at least trying to keep the figures quiet until it is known that the second tender is not required. The other traditional practice of sipping liquid is still observed at some tender openings.

The immediate action required is to tabulate the tenders in ascending order of amounts, giving the contract period offered if this is not firmly stipulated in the documents. If any tender has been qualified, either on its face or in its covering letter, or if any space on the form of tender has been left uncompleted, these facts should be recorded. The completed schedule should be signed by the persons present, or by at least two of them if there is a larger number. After any photo-copies have been taken the original schedules and forms of tender should be passed to the client or be kept on his behalf by the architect.

If a tender is received after the stipulated date and time, it is desirable to reject it even if it is the lowest. The argument for this is that the firm may have hung back until it heard on the 'grape vine' and then took an unfair opportunity to undercut the other tenderers. A client or adviser inviting tenderers frequently will benefit in the long run by maintaining a reasonable discipline. It is, however, the client to whom the tender is made, and if he chooses to overrule this policy, he may of course do so and ask for the tender to be considered. The position is in any case different when the firm puts in the written tender late, but rings through its figure before the opening—a valid offer has been made, even though oral.

The bills of quantities of the apparently acceptable tender should be obtained for checking technically as soon as possible. Usually the tenderer will be the lowest in price, but he may be one offering a shorter contract period or one who does not qualify his tender adversely. If tenderers were each sent two copies of the bills, the one sent in at the same time as the tender should be taken from the sealed envelope bearing the firm's name; those of other tenderers will be returned unopened as soon as it becomes clear that they will not be needed. If tenderers were sent only one copy each then the firm being considered should be sent a further copy into which to copy its prices as soon as possible.

If tendering is very close, and time is short, it may be desirable to ask more than one tenderer to forward bills, in case an error affects the order of the running. This permits the arithmetical stage of checking to be done in each case, but no attempt should be made to put the two sets of prices alongside each other: only the one still in favour should then be examined. Since there is always the feeling that information may be misused, however innocently, this expedient should be adopted only in pressing circumstances and then only with the consent of the firms concerned.

If tenders have resulted from open tendering there is the possibility that the firm immediately under consideration may be unknown to the client or his advisers. This means that an immediate check should be made on the criteria outlined in Chapter 6. This should be done while the tender is being checked technically but it may nevertheless lead to an overall delay. The most important points should be pursued most rapidly, so that if the tender is to be rejected the next most favourable can be taken up quickly.

Whether open or selective tendering has been used, it is desirable to ask the tenderer being considered to furnish a list of proposed staff and an outline programme as soon as possible. Neither of these will form part of the contract, and the contractor can change their details, but they are an earnest of good intent and may throw up some important point.

At this early stage the tenderers should be given as fair an indication as possible of where they stand. The most favoured tenderer will know his position from the approaches he receives. The next two should be told that they are being held in reserve against the possibility of the first withdrawing. Any other tenders are best declined outright; since firms may have tenders under consideration for more jobs than they can undertake it is right to remove doubts as soon as possible. But no details of amounts or order ought to be given at this stage; to do so would be to invite difficulties through firms knowing the margins between themselves and the next one up on the list. (They may have their own ways of finding out, but this is another matter.) Once a tender is accepted all firms should receive a list giving the names of tenderers in alphabetical order and of the amounts tendered in numerical order. This helps each firm to check the keenness of its own tender without competitors' tenders being identifiable. Any list published in the press with firms and figures co-related must be published only with the consent of all the firms involved.

Alternative tenders

If alternative tenders were invited initially or were allowed during the tendering period, a decision must be made on whether each alternative need be examined in detail or whether a comparison of the overall tender figures or other parts of the offers has narrowed the field to one alternative without more ado. No special problems of procedure arise in examining more than one alternative, although this may keep two tenderers in the running for some time owing to a reversal of their order as between the alternatives.

The position may be more complex where the otherwise eligible tenderer has made any qualifications of his own to the tender in a covering letter. These qualifications may affect things like the contract period or the phasing of sections of the work or may consist of a refusal to give a firm price tender. Strictly speaking, these qualifications may be ignored if not notified when putting in the tender; a client can accept an unqualified offer in a form of tender the moment he receives it, if he so chooses, bills or no bills. Any letter sent with the tender must be considered however, even if it leads to the rejection of the tender. If the letter arrives with the priced bills it is possible to stand on one's dignity and pass on to the next tenderer but it seems better to try to reach a solution. In most cases, and whenever the letter arrived, the tender as originally asked for should be considered. If that has not been submitted but only a modified version, then the original version should be requested. This will mean that the parity of tendering is observed and like is being compared with like in the tender list.

It sometimes happens that a tenderer will not give a tender on the original basis, particularly over the question of a fixed price or if he cannot meet the completion date. It would be far better in such circumstances for him to say so on receipt of the tendering documents. However, if he does not say so the choice will be between rejecting him outright and passing to the next tenderer or else asking the next tenderer or two for revised tenders based on the amended terms offered. If the client or the architect is not prepared to consider the amendment the problem resolves itself at once; if they are, then the better course of action is to ask for revised tenders from other tenderers, as just suggested. This not only maintains parity but also clears up any doubt about the true order of tenders. The omission of an allowance for fluctuations can sometimes work wonders in this respect.

Examination of the priced bills

Even before the priced bills of the favoured tenderer are examined the quantity surveyor will have compared the total tender sum with the others and with his own estimate, to see whether the differential is unusual. It is a healthy pattern to find, say, three firms out of six well within a five per cent bracket for straightforward work. If firms straggle up to ten per cent above the lowest tender they probably only want the job at a good price, if at all. It is the firm on its own some way below the others that will alert the quantity surveyor at once. It is relevant that a given percentage margin on the whole tender represents a significantly larger one on the sections priced by the firm—ie, excluding PC and provisional sums. But if the firm is a normally reliable firm its bills should be examined before any misgiving is expressed.

As soon as the bills are available some superficial points can be checked. If there have been any amendments circulated during tendering the bills should show any altered quantity or description. There may be some qualifications notified to tenderers which cannot be checked with certainty in this way and these should be noted. The total of the bills should be compared with the figure on the form of tender; the latter is the offer and will become the contract sum.

THE STRUCTURE OF THE PRICING

As the technicalities of the pricing are approached a start is best made by analysing its overall structure. It may be that the cost plan is in a suitable form for making a comparison; if so it will be one unbiased by the sight of the actual tender details. A main consideration is whether preliminaries have been priced separately at anything like an economic level or whether they have been absorbed into the measured prices, either evenly or in particular places; the level of measured prices to be expected will obviously depend on this. From the client's point of view, preliminaries spread out evenly instead of being priced separately mean that the setting up of the site organisation is paid for progressively and not all in the early certificates.

The measured prices themselves can also be reviewed broadly at this early juncture, to see whether particular sections of the bills have been priced high or low. High prices may be due to an allowance for preliminaries but some firms deliberately price the early trades high so that the job is partly financed from early payments; later trades will then be pared to keep the total tender at the right figure. Apart from the client having to commit his finances earlier, this practice can also lead to a very adverse position if the contractor becomes insolvent just after carrying out this highly priced work and his employment is determined at that stage. His successor's prices will probably be higher for the later trades in any case, but this manipulation will aggravate the gap.

It is always going a little near to the ethical bone to look for high and low prices in considering the effect of possible variations, but if a bill of reductions is going to be inevitable the pricing relationships will need to be considered as soon as possible to see that the changes being proposed are going to achieve their object. As the tenderer will be advised later—but before acceptance—of what is in the air he will have the opportunity to give voice to his objections if he has any.

THE DETAILS OF THE PRICING

A contractor's prices are his own, both before and after acceptance, but they must be considered carefully all the same. The main aim is to look for any obvious mistakes in individual prices and not for differences that are merely matters of opinion. Somewhere below the level of manipulation considered under the last heading there is still plenty of room for the totals of particular trades or work sections to vary between tenderers (and certainly by comparison with the quantity surveyor's opinion) by up to 10 per cent even when the totals of tenders are within 2 per cent. To be sure that an error has been found at the level of item prices, the quantity surveyor must be looking for something that is wide of the 'average' by at least 25 per cent in many cases. Anything less than this margin may be due to differences in efficiency of working or buying or to the method of allowing for plant or scaffolding.

Items priced in relation to the incorrect unit of measurement are likely culprits: typical examples are superficial instead of cubic, kilogram where the surveyor has billed by megagram to avoid astronomic figures. The misplacing of the decimal point in the price is another potential error to keep in mind. It is the items 'where the money is' that should receive the most attention. Differences on small value items, such as labours, are often quite considerable due to the lack of feed-back data from site; it must be remembered that these items usually have a small influence on the overall tender.

Inconsistent pricing of several related items, such as beds differing only in thickness, can niggle the tidy mind of the quantity surveyor but the temptation to ask the tenderer to tidy these up should be resisted unless pro rata prices are specially likely to occur. It is perhaps wise to note only cases where the prices have been reversed or the differences are extreme.

A complete arithmetical check must be put across the whole of the bills, since anything found here must be an error. This will include looking for unpriced items, other than those deliberately marked 'Nil' or 'Inc'. Regular sources of error are in omitting to transfer a page total to a collection, or in transferring the wrong amount. Any errors found should be summarised by item reference, using double cash paper, and the net total discrepancy found. Any pricing queries are conveniently listed in a similar way, except that any discrepancy calculated will be an assessment only. A note of any unauthorised amendments made in the text of the bills by the tenderer completes the examination of the body of the bills.

There remain in the issued documents any schedules and appendices that the tenderer had to complete, covering aspects like dayworks and fluctuations. These should be checked for completeness and possible error. If they were not bound into the bills the quantity surveyor must first make sure that all the pages have actually been returned; an oversight at this stage is all too easy, especially if an unusual schedule was issued.

Action with the tenderer

The quantity surveyor is now ready to meet the tenderer, unless there are no points to raise with him at all. Two considerations need to be borne in mind.

In the first place the quantity surveyor is not here engaged in the position that he will occupy when acting between the parties to the contract during the progress of the work. He is concerned to see that his client is entering into a sound and commercially favourable agreement, though his dealing will still be purely professional. The tenderer, it is to be assumed, is quite capable of seeing that his own interests are secured. Neither party is yet committed, except perhaps by the need of work or the pressure of time.

In the second place, it is not the quantity surveyor's function at the meeting to discuss with the tenderer the overall checks that have been made or any views on the structure of the pricing—which is the tenderer's response to the bills with which he has been presented. The bills as priced are the bed on which the contractor and the client must both lie, in whatever harmony is achieved. The quantity surveyor must concern himself with the bills as they are, and discuss only the details with the tenderer. But he may, by judicious snatches of conversation, be able to satisfy himself as to the philosophy of the pricing—and perhaps even about the motives behind it.

THE NORMAL PROCEDURE

The queries picked up in checking through the bills will be raised with the tenderer, any prices being either confirmed or altered and any arithmetical errors being corrected. Any alterations to the words of the text should be resolved, usually by removing them and if necessary changing the unit prices to represent the original. The architect may be prepared to concede some change; unless this is of major calibre it should not be regarded as affecting parity. Any alteration of quantity made by the tenderer should certainly be cancelled and the original held to. If the tenderer turns out to be right the adjustment will become a variation. If, owing to shortage of time, the detailed prices for a sub-trade were not inserted in the priced bills originally put forward, they must now be inserted by the tenderer himself and then be checked in the same way as the rest. The resultant amendments to the arithmetic of the bills are now ready to be made right through to the final total. This figure must not be altered unless it differs from that on the form of tender, which must prevail.

There will now be a discrepancy in the figures on the general summary. This may be dealt with by inserting an addition or omission item in the summary immediately before the final total, to take up the difference. The item should be worded as being a percentage adjustment to all prices in the bills, except prime cost and provisional sums, which will apply therefore to all variations. It is recommended in the Code of Procedure for Single Stage Tendering that preliminaries should also be excluded from the adjustment but, with respect, this does not appear particularly logical; preliminaries are the subject of competition and may in any case be lost in the measured prices. The fact that preliminaries are not always adjusted in the final account does not seem particularly relevant.

While this method of absorbing a discrepancy is the most rational and is the best for any fairly large correction, it is often felt to be pedantic when used for dealing with small sums. Sometimes one or two prices with suitable quantities can be used to take up the difference or, even more easily, a lump sum adjustment can be made to 'water and insurances' in the general summary or to an item in the preliminaries. Any gross distortion of the original prices should be avoided. Whatever the method employed, the tenderer's agreement will obviously be necessary.

If daywork was priced in the bills and reflected as a sum in the competition it will have been considered by the quantity surveyor in what has gone before. If, however, rates and percentages have been set out in a schedule without any effect on the tender, the tenderer may have been tempted to pitch them high. It is quite in order to negotiate on this non-competitive section and to try to secure a reduction. This may be difficult, since the tender is hardly likely to be rejected on the strength of the daywork terms. The moral is that daywork should be quantified, put into the bills, and treated as a contingency.

Where formula fluctuations are to apply in the contract there is nothing to do, other than to prepare the schedule (if not already in the bills), analysing the bills in accordance with the formula rules, and to be clear that the correct 'base month' has been used if the date of tender has been moved, as mentioned as the beginning of this chapter. Where fluctuations are to be dealt with by the traditional method, there remains the appendix giving the basic prices of materials—if market and statutory fluctuations apply—or simply a list of materials if only statutory fluctuations apply. The tenderer should be asked to delete any items which were not stipulated, unless he has very cogent reasons for leaving them in. If the basic prices were already printed in little more can be done, as the tenderer has not fixed his own basis. Where sources of supply can lead to price variation he should be asked to declare his intended sources and these should be embodied in the appendix. Doing this will prevent any changes being made simply on account of a change of supplier but may still leave some adjustments to be made without any market fluctuation, where the basic prices have been supplied to tenderers. If the basic prices have been inserted by the tenderer himself he must be prepared to substantiate that they are 'market prices' so that any future adjustments will be due to genuine market fluctuations. Here he will often need to produce quotations and these should become part of the appendix as establishing the intended source.

When all is tidied—and this may take several meetings if things are problematic—the tenderer should be asked to initial every alteration made to the bills as he sent them in. Only where he completes any blank spaces may this be dispensed with. There are sometimes matters over which the tenderer will need to send in a confirming letter. It is best here for him to agree the wording with the quantity surveyor in advance so that the intentions of both are clearly represented. This letter should be included in the eventual contract appendices unless its contents are reflected in other incorporations to the documents.

MAJOR ERRORS IN TENDERS
It sometimes comes about that examination of the priced bills or perhaps the copying of them by the tenderer reveals some substantial error. If the error is a high price or an over-calcuation in the arithmetic the effect will be that the tender sum has been inflated. Since this inflated sum is still the most favourable obtained in competition—otherwise this particular tenderer would not be considered—the tenderer is quite in order in asking for the tender to be maintained and the error adjusted like any other. The quantity surveyor should be prepared to recommend this, although he can always explore the situation to see whether the tenderer is feeling generous.

If the error is in the other direction things are more delicate because the tenderer or the client may have second thoughts. The tenderer should be notified first, and at once, and be asked whether he wishes to stand by his tender or withdraw it. If he does one of these things the position is clear, although the quantity surveyor must consider carefully whether, if the tender is not withdrawn, it is likely to be too low to form the basis of a steady contract and whether its rejection should be recommended for this reason. But the tenderer may ask for an amendment to his amount, particularly since his tender is probably well below the next and he will by now either know or guess this. If the requested amendment is large enough it may change the order of tendering and, again, the line of action is clear, all else being equal. If however, the amendment merely narrows the gap, opinions are divided on what is best to be done. Some would say that such an amendment should never be countenanced, on the ground that discipline among tenderers ought to be maintained. This is a point of view that will appeal to a large client, conducting frequent dealings with the same group of firms. The client making only a rare capital investment in building is not likely to see matters in quite the same light and it is hard commercial sense for him to do otherwise.

In all cases, therefore, it is prudent to notify the client of the full facts before proceeding further with the tenderer, giving him advice on action but letting him come to his decision. In no case is the quantity surveyor, or the architect, entitled to reject the amended offer on his own authority. If the client decides to consider the modified offer then the examination will proceed. The narrowing of the gap between the most favourable two tenders may be such that it is now advisable to examine both. The next firm in order may well have an error which they are prepared to adjust and which brings them into first place. If this happens it will have a quite salutary effect on the other firm and is quite a correct procedure in the circumstances, which are now verging on negotiation. What must be watched for is that any major error in the tender is genuine. If the tenderer reports the error before sending in his priced bills, the way in which it occurred must be shown to be bona fide beyond all doubt and this will mean going back to the original copy and to other data. If any doubt remains, rejection should be advised. Even if the error comes to light in examination, nothing should be assumed too lightly; it is not unknown for the imprints of the correct figures to be found below the 'error'.

If in the end a client accepts a modified tender against the advice given to him, the architect can always exercise his own discipline in the future. He does not have to have a dubious firm on his selected lists.

The quantity surveyor's report

When all points have been agreed with the tenderer, with or without the hiatus of a major error and an early report, the quantity surveyor is ready to let the architect have his routine report. While this is written as from one technical person to another, it should be worded so that the salient points made will be clear to the client, if it is necessary for him to be aware of them. Needless to say, he will not be very interested in the minutae of an 'extra over'; probably the architect will not enthuse greatly either.

Most comments should therefore be fairly general. If the tender is high or low this should be stated, with any reasons that can sensibly be given. Any peculiarities in the pricing that distributes the money unevenly within the whole are also important commercial aspects. If any amendment to the tender sum, or to some such condition as the contract period, has arisen since receipt of tenders, or if some qualification has been made, these matters should be explicitly stated along with information as to whose authority was obtained for proceeding with the examination of the amended terms. Small issues in such realms as the basic price list are the quantity surveyor's domestic concern and should not be raised now. All that is needed is a statement that all minor items have been cleared. It may be advisable to send the report in early to save time and to say that it is subject to confirmation of a few small matters. Clearance of these can be notified later, or even be assumed to have occurred in the absence of any other letter from the quantity surveyor.

While the report should not descend to the trivial it must be positive. Acceptance from the commercial point of view should be recommended or otherwise. If there must be qualifications or the statement of alternatives the issues must not be blurred. Decisions have to be made which are based at least partly on the report; the quantity surveyor will have to stand or fall by it although he should not be expected, in retrospect, to have been a crystal-gazer or a detective.

Even though the tender as a whole is within acceptable limits, the client may need to know how it is divided at this early stage. His funds may be coming from several sources and be earmarked for particular sections of the work, or between some such conceptual entities as 'net' and 'gross' costs. An analysis of the tender in these terms is best given as a separate statement but should accompany the report to assure the client that all is well. In simple cases the information can be asked for on the form of tender but may be affected by errors or the apportionment of preliminaries.

Here the quantity surveyor ends his activity until he becomes involved in the final pre-contract activities outlined in Chapter 14. Unless, that is, a reduction bill is needed—but this is hardly the note on which to end a chapter.

CHAPTER 12

Examination of tenders not based on bills of firm quantities

When contract arrangements are not related to bills of firm quantities the examination of tenders will pass through the same main stages, and be conditioned by the same principles, as were discussed in the previous chapter. The quantity surveyor will still be looking for parity between tenders, coupled with a commercially satisfying deal for his client, obtained in a professional manner. There will be differences of detail, and these are picked out under the headings that follow starting with the forms closest to firm quantities.

Bills of approximate quantities

The main difference here will be in the quantity surveyor's outlook during examination, which will depend on how approximate the quantities are. If they are a close comprehensive reflection of a well developed design then no difference may be necessary; the further the bills stray from this position, towards becoming an arbitrarily weighted schedule, the more will the surveyor temper his approach.

While relatively high or low priced trades should be looked for in firm quantities, their effect is fixed in the quantities as given, unless substantial variations upset the balance. It is this disturbance of balance that becomes the bogey with approximate quantities. If, therefore, trades are unusually priced the quantities for those trades should be reassessed as carefully as time and information permit, to see what impact the prices are likely to have in the reassessment. Since the surveyor will in effect be calculating a private revision to the tender sum it should be borne in mind that the next tender may exhibit just the opposite pricing oddities and that a revision of it might produce a revision of the original order of tendering. Thus two tenders are likely to be examined in the case of approximate quantities where the apparent gap is proportionately greater than in the case of firm quantities. This is a function of the uncertainty in the approximations. In most cases an operation like this will be unnecessary or impracticable but the possibility should not be forgotten.

The other difference in emphasis will be in reviewing the consistency of pricing within trades. Not only are the quantities approximate but so to some degree will be the selection of items in the bills. The possibility of more pro rata prices in the final account is thus greater. In important cases it may be reasonable to ask the tenderer to tidy up the relativities of his prices to ease future work, although he is not obliged to do so. The total of the tender should still be maintained unaltered if this is done.

It may be desirable, if the design work has taken any big steps during tendering, to negotiate rates for fresh items, or items which have changed drastically in quantity. This work is often done with less trouble before acceptance and the fresh items can be appended to the bills in a schedule of rates. Otherwise approximate quantities can be reported on just like their firm cousins.

Schedules of rates

Tenders based on schedules of rates present the same problems as those based on approximate quantities, only more so. This is because schedules of rates are most commonly employed where the design uncertainty is too great to allow any quantities to be sent to tender. The position is somewhat easier where a pre-priced schedule has been used.

However, except in the unlikely case of one tenderer quoting consistently more favourable percentages 'on or off' the great bulk of the sections in the schedule, the nettle of quantities will now have to be grasped. It is likely that all tenders will have to be considered in this way unless any are consistently unfavourable.

In fact, before the tenders are received the nettle should already have been grasped, to save time. An estimate of the work should have been prepared, priced at schedule rates and presented as a series of lump sums corresponding to the sections of the schedule on which distinct percentages may be quoted. Little quantity detail need have been assumed, so long as the balance in money values is reasonable. The early preparation of this estimate also helps to remove any suggestion of bias in preparing it when the percentages are known.

Each tenderer's percentages will now be applied to these values and any lump sum for preliminaries added, if they are not also a percentage, and a total competitive figure will thus be ascertained. This will set the tenders in order, as is shown in the simplified example given in Figure 8. From here on the procedure follows that for bills of approximate quantities, with a consideration of high and low pricing and the effect that a swing in the proportions of items could have. No irregularities within sections of the schedules need be considered, since the pricing differentials have already been fixed and imposed on all tenderers. The range of items given is also likely to be more than is required in most cases, not less.

The action with the tenderer is thus likely to be minimal and the whole examination may be carried through quite quickly. It is desirable to give an interim report on the order of tendering, as soon as the quantity surveyor's weighted estimate has had the percentages applied to it. Until this is done everyone will be in the dark as to where things are heading; the client will have little idea of his commitment and none of the tenderers can assume that he is in or out of the running.

The schedule that is not pre-priced is harder to assess. If quantities are to be used they must show at least the main items of value and be priced for each tenderer, since there are only net rates in the tender and no percentages. The only alternative is to make a superficial comparison of the levels of pricing of all tenders; this is as unreliable as pricing quantities is laborious. Thus blank schedules are hard to examine, although they are fairer to tenderers by not forcing price differentials on them.

Prime cost

Tenders for prime cost contracts may be considered as both the easiest and the hardest to assess. They are easiest in that all that can be done is to assess the 'plus' element. If this is on a fixed fee basis the various tenders have simply to be listed and compared as for a lump sum contract. If it is on a percentage addition basis the several percentages are applied to the estimated labour, materials and plant contents and again the totals are compared. There can be no hidden innuendoes and any mistakes will be easily seen.

These tenders are the hardest to assess because of the difficulty of being sure of the estimate of the prime cost element. Often the extent of the work is uncertain and so, therefore, is the estimate on this score, but this problem does not really affect the comparison of tenders. These are being compared on a proportionate basis and any change in the estimate will affect all tenders in the same way. Where the trouble lies is that the various firms may have different levels of efficiency, and consequently of cost, in performing the work, particularly in respect of labour and plant. This is not reflected in the tendered fee but is payable in the final account. Indeed the fee also rises with the cost if it is assessed on a percentage basis. It is true that where the JCT Fixed Fee Form is used there is provision for stating the estimated prime cost of the work but there is no effective mechanism in that contract for dealing with any excess incurred (see *BCPG*, Chapter 24). Where some form of target cost is employed the problem remains the same although it is somewhat eased in scale by the bonus and penalty arrangements which damp down the effects on the client of the contractor's efficiency or inefficiency.

In the end, then, the assessment will be affected—at least as much as by anything else—by the architect's or quantity surveyor's previous experience of the firms on similar contracts. This is why negotiation is often adopted

SCHEDULE	ESTIMATE	FIRM A			FIRM B			FIRM C		
Section	Total	Order	Percentage	Total	Order	Percentage	Total	Order	Percentage	Total
	£			£			£			£
Excavation	4,000	2	+15	4,600	1	+10	4,400	3	+20	4,800
Concrete	16,000	2	+2½	16,400	3	+4	16,640	1	-5	15,200
Brickwork	7,000	1	+3	7,210	2	+5	7,350	3	+7	7,490
Joinery	3,000	3	+20	3,600	1	+12	3,360	2	+17	3,510
Painting	2,000	1	+8	2,160	2	+15	2,300	3	+20	2,400
				33,970			34,050			33,400
Water and insurance		2	(1%)	340	3	(1½%)	500	1	(½%)	170
Preliminaries (sum)		1		2,300	3		3,500	2		2,650
				£36,610			£38,050			£36,220

Figure 8: Comparison of schedule of rates tenders. 'Order' in columns two, three and four indicates the relative competitiveness of the section or the like. No tenderer has a clear advantage on a first viewing on this basis.

for prime cost work, the only advantage of competition being that it may depress all firms' fees somewhat.

Drawings and specification

Where lump sum tenders are obtained without quantities that will form part of the contract, the comparison of tenders should be a straightforward case of 'like with like', unless tenderers have qualified their tenders. The tenders may vary more widely than when there are quantities, both because tenders of small value tend to do this anyway and also because there is competition in both pricing and the accuracy of the quantities. The firm with the scantiest quantities has a head start in the race. The positive discovery by the quantity surveyor of a tenderer's error is very difficult and he can gauge the overall tender only in relation to its competitors and to his own estimate, unless a breakdown of tenders is obtained in the second of the two forms considered below. If daywork terms were requested from the tenderer, these will inevitably be in an appendix and will not have affected the level of the lump sum and so must be inspected separately and, if needs be, queried.

The main reason for obtaining a breakdown is for it to serve as a basis for variations and it should not be obtained at all unless the size of the contract really warrants a pre-determined basis. It may take one of two forms. In the more simple cases it may consist of several lump sums adding up to the whole; the divisions required will have been set out in a schedule accompanying the enquiry and returned by all tenderers. The analysis will be used as its stands.

In the more complex cases the favoured tenderer only is asked for an analysis in the form of the priced quantities used in arriving at his tender. When such a breakdown is obtained it should be scanned for any major omission. It should be remembered, however, that the estimating may be done on a very inclusive basis and even fairly important items shown in bills of quantities may not show at all here. No detailed checking is called for, since the quantities are not to form part of the contract. When the analysis has been examined and found reasonable for its purpose all quantities and extensions should be deleted and the item descriptions and prices alone become a schedule of rates—solely for the purpose of variations. Any error in the now removed quantities is entirely at the tenderer's risk. If the document is too untidy, or the descriptions too vague, it can be redrafted to suit its new purpose. In any case, the tenderer should agree to, and sign, the schedule as prepared.

All-in service

If an all-in service is made competitive it is fair to do this only to the stage of outline designs. Some of these designs will be eliminated by the client and the architect without the need to go into their economics. If more than one remains the quantity surveyor may be required to go through a cost comparison; even one design will warrant a cost assessment. In each case the methods of normal cost planning are appropriate, although obviously this will mean a cost plan for each design if the layouts vary or if radical differences are introduced into elements like the structure or the services. The aim will be to assess the construction price level put forward by each tenderer in relation to his own design, rather than to compare prices directly. The primary emphasis in selecting the contractor is thus on the design—but with this emphasis tempered by the knowledge of how economically the tenderer can produce that design. This exercise by the surveyor may be paralleled by another to assess the schemes on the basis of costs-in-use.

With the two variables of design and cost thrown into competition it is quite in order for a measure of negotiation to go on at the eliminating stage. In any case, it may be that the probing which goes on into the data submitted by the various tenderers will be difficult to distinguish from negotiation, as it is likely to produce voluntary modifications of the outline offer. Since negotiation is going to continue with the tenderer eventually isolated to produce a contract figure, it can well be initiated by the quantity surveyor while the spur of competition remains, to ensure a good basis for the subsequent non-competitive negotiations. Even when this first stage is over, and only one firm is being dealt with, it is desirable for one or two other firms still to be in reserve and be known to be so. Negotiations are then more likely to proceed evenly and if they do break down one of the reserve firms can be recalled. The main difficulty here is likely to be the lapse of time before this situation emerges, with the possibility that the reserve firms may no longer be interested or that the client's programme may be prejudiced.

At a fairly advanced stage a basis for variations and a method for interim valuations will need to be evolved, using the data built up during negotiations. The contract will be arranged on a 'drawings and specification' basis and only if it is of an unusual size for this treatment may there perhaps be some difficulty here.

CHAPTER 13

Principles in negotiating for tenders

If there is any part of the subject of this volume about which it must be said that experience counts it is the matter of negotiations. The present chapter is thus significantly brief and all this is intended to do is to set down some principles to be followed as far as possible. They apply most widely where bills of quantities are used but many points will be applicable to other contract arrangements.

The pattern of negotiations has already been considered in part in Chapter 6, since it is allied to tendering procedure. This is by no means a tidy division, as negotiation is all of a piece. Its only justification is that negotiations span both the pre-tender stage and the receipt of tenders and the division made puts a share of the subject into each relevant part of the book.

The authority of the quantity surveyor
For negotiations in any sphere of life to be satisfactory the parties to it should be equally matched. They should enter into it freely, with equal facilities and mutual respect, and with the possibility of calling the whole thing off if they are not satisfied. Seldom do such conditions of idyllic bliss obtain and this is true of building negotiations in particular.

The quantity surveyor finds himself in a position where he is not a party to the negotiations but is conducting them on behalf of one of the parties. This is important; as was said in a previous chapter, in the examination of a competitive tender the quantity surveyor is not assuming a quasi-judicial position but is acting for the client. For him to act effectively he should be able to set out the financial stall himself and not be brought in when either the client or the architect has done some discussing and perhaps decked the stall out in a quite inconvenient way. He should also seek a clear authority to carry negotiations to completion, with the twin provisos that he is not obliged to settle come what may and that all he is doing is obtaining an offer which the client alone may accept, if he wishes.

For the surveyor and the contractor to start operations together is likely to produce the most harmonious relationship, but the persons doing the work should also be as evenly matched as possible for their own stages of the proceedings. There will often be a detailed stage followed by a more comprehensive stage in which all the points of difference are set side by side

136

and agreement is reached. Persons of equable temperament, who are able to see the wood as well as the trees, make the best negotiators. While they should not be weak they should remember that it never pays to prove the other person wrong in too aggressive a manner. It is better to allow him to extricate himself gracefully and concede a point by bringing some fresh element into the reckoning. It is also desirable to proceed in a frank atmosphere, rather than to create an air of mystery as though securing some undisclosed upper hand.

The data for negotiation

There are two broad ways of obtaining data for the negotiation: to start with the prices of some other recent and similar project, obtained in competition, or to go back to first principles. There is something to be said for each.

If another project is selected for comparison it must obviously be one priced by the contractor and preferably it should be one obtained in competition, to establish a firm base line for the pricing in it. If the quantity surveyor did not himself deal with that job he should, with the contractor's agreement, obtain leading details of it. These will include particular obligations imposed by the contract, information about the site and the programme, and a list of the tenders received. In addition, the contractor himself should be prepared to provide at least in outline, an analysis of how he distributed major costs and charges in his pricing. The nearer this comes to the data provided in the second approach outlined below the better it will be. The aim is to establish a structure of pricing and this it is the prerogative of the contractor to set out, unless he is clearly distorting it, and always with the quantity surveyor able to offer constructive suggestions on how it may be improved to fit the job in question.

The advantage of this approach is that it is close to market conditions as a source. The less similar the previous job is to the present one, the less value will the approach have. At some stage across the spectrum of projects it becomes better to rely on first principles, while still referring to market conditions to verify what is being proposed.

In this second approach not only the structure of the pricing but the levels of the individual parts must be decided upon, since there is no other job to break down into this detail. Site supervision, major plant and scaffolding are items which must be considered in total before their allocation in the pricing can proceed. On the other hand, head office establishment charges, routine insurance and profit need to be arrived at as percentages to be applied to whatever is otherwise calculated, and therefore to vary with the value. These and other considerations will be applied differently in considering the contractor's own directly measured work and daywork, sub-let trades, and PC work.

The course of negotiation

How and when the data of negotiation are ascertained and employed will depend upon the pattern of negotiation. But whenever this activity takes place there is much to be said for the contractor and the surveyor working together and sifting and applying the data together. By doing this they should develop a better working relationship and secure a deeper appreciation of the job in question. It also means that a series of smaller issues can be hammered out, rather than one person waiting for the other to put forward a complete tentative offer with prices quoted high (or low) on the assumption that there will be an inevitable knocking down or pushing up—a much more drastic operation than a progressive adjustment by the two negotiators. But even where this latter method is used it should be clearly understood between the two that any agreements over parts are provisional, pending the sight of some larger part, or perhaps of the whole. Neither should go back lightly to make frivolous changes but the necessity to extract costs from one place and put them more rationally in another can arise and sometimes assumptions or concessions may be unwarranted when viewed with hindsight.

The quantity surveyor in particular must be prepared for the possibility of higher pricing than in competition. The character of the job that has led to the choice of negotiation may make higher pricing inevitable. But the 'extra over competition' element does not lie here; it is partly in the fact that discussion may have disclosed what is often unforeseen in competition and then becomes the subject of a claim or loss. To this degree the negotiation should not be increasing the final capital cost to the client. The extra may also lie in the initial choice of a single contractor on grounds other than price; he is presumably considered better in some way than other contractors and this must be paid for.

In the end a contractor taking part in negotiation knows how things stand and that he may secure a better profit margin than usual. There is nothing untoward in this; many will hold that the industry as a whole has margins which are far too low. But because of this natural tendency there must be the possibility of breaking off negotiation that is heading above what is reasonable. And the need for comparisons also explains why negotiation can never be considered as entirely replacing competition.

The problem of a deadlock over price level highlights the reason for the quantity surveyor being in the negotiation from the beginning, rather than arriving like an ambulance man when injury has occurred. It also explains why neither party to the negotiation should be committed to contract without escape until a sufficient and satisfactory stage has been reached. It may be necessary, due to pressures of time, to finalise many details after entering the contract but they should flow fairly painlessly from the contract basis.

Reporting the negotiated offer

When negotiations are completed to whatever stage is to serve as the contract basis the contractor should put forward a firm offer embodying the results. The quantity surveyor should report on this in the same way as on an offer obtained by competitive tendering, although not all features of that report will apply here. Since there is no list of tenders the quantity surveyor should include in his report an assessment of how the offer obtained compares with what competition would have yielded, and any reasons for any part of the difference. The client may be prepared to pay more for the advantages accruing from negotiation but it is only fair to him that he should know how much more he is being asked to pay.

CHAPTER 14

Some final pre-contract activities

When competition or negotiation has run its course and come to an end the client will be faced with an acceptable offer, recommended by his professional advisers, from the one who is about to become the contractor. As soon as he has issued a letter of acceptance, or the architect has done so on his behalf, there will be a binding contract even though the formalities are yet to be executed. A letter of intent, on the other hand, binds no one but is a useful way of signifying the stage reached. If parties take action without written evidence of an agreement there may still be a contract implied out of their action or condonation of the other party's action.

As a general rule the aim should be to issue a letter of acceptance within two months of the offer, so that the firm is not left in uncertainty. The clearing of statutory approvals and arranging of funds may militate against this but the tenderer always has the option of withdrawing his offer at any time before acceptance.

There remain several matters that have to be tidied up. Some may have been cleared during earlier discussion but they may be dealt with in the formal packets set out below.

The pre-contract meeting
Particularly on large projects, a meeting will be held under the chairmanship of the architect to ensure that the various ends have all be tied up. As well as the client and the contractor-designate, the quantity surveyor and any consultants should be present.

The dates for possession of the site and completion of the work should be confirmed or modified in the light of any developments since tenders were invited. (Changes likely to have any financial effects will already have been evaluated between the parties.) In the context of these overall dates the contractor should be asked to present his programme of work and a method statement. So far as the physical side of the work is concerned these items are his own responsibility and at no stage should they be approved in such a way as to change this position. But there may be, particularly with alterations or phased work, points to which objections may be raised and it is far better to clear the air before the parties are committed; at the first site meeting it may be too late. Another aspect is that of the supply, by the design team to the contractor, of details, schedules and the like; as precise a programme as can be drawn up at this early stage should be established as a basis, for review as work progresses and to guide the team on its order of working.

The firms in mind for nominated work, so far as they are known, and if not already named in the tendering documents, can conveniently be named, with any available information about their programmes; under the JCT forms at least, a contractor has certain rights of objection and it is better for these to be voiced when the position is still contractually fluid. There may also be nominated work for which the contractor will wish to tender and the architect should now indicate those cases to which he agrees. Since sub-letting is regulated under the standard contracts, the contractor should be encouraged to put forward (when the bills do not) the firms he proposes to use or from among which he will choose. He need not commit himself but it is better to clear his way ahead or to block it entirely, than to leave the matter unresolved and possibly a source of argument and delay later.

Several of the foregoing points affect the details covered in the appendix to the contract, or its equivalent. All these details should be confirmed, since the tender documents will often say that the details forecast in them are subject to confirmation.

Another important area of discussion, not highlighted by the appendix, is that of the insurance requirements. While the standard contracts give their own varying but specific conditions on these matters, it is often the quantity surveyor's preliminaries that enshrine the extent of the insurance required against the various risks. He should therefore have the information to hand, along with a note of who is responsible for effecting the insurance in each case.

Preparation of the contract documents
The exact work done by the quantity surveyor in the drawing up of the contract documents will vary with the contract arrangements but since he will usually have drafted the preliminaries he is definitely the one best placed to prepare the contract conditions and appendices. Any deletions of, or amendments to, any of the conditions must be made physically in the document; it is not sufficient to rely on the wording in the preliminaries, which is merely a declaration of intention. Preferably the parties should initial the amendments also.

Where there are bills of quantities the surveyor will certainly be responsible for these. Where appearance is paramount there may be a specially bound copy for contract purposes but at least a fair copy of the prices will be made in such cases. It is often satisfactory, however, to produce the priced copies by photo-printing; several copies may be produced for those entitled to receive them with the certainty that there will be no transcribing mistakes and with the advantage, as it often is, of enabling everybody to see any changes that have been made during the pre-contract stages.

Execution of the contract

The contract may be executed under hand (in the proper legal sense of the term, of course) or under seal. The usual factors conditioning the choice between the two methods are whether it is desired to secure six or twelve years' protection for the client under the Limitation Act and whether the client's status requires him to contract under seal, although this is seldom necessary today. A contract in which there is no consideration from the client's side must also be under seal. In Scotland the facility of contracting under seal is not available: there are other ways of securing the advantages that it offers, while a gratuitous contract can be valid as there is no doctrine of consideration (see *BCPG* Chapters 3 and 26).

The actual question of *when* the contract is executed is of less moment. Often enough it is not done until after work has started and when the initial excitement has tailed off. A letter of acceptance, or even oral acceptance, is more likely to set up the legal relationship between client and contractor and the timing of this should never be so early as to prejudice the successful conclusion of any outstanding negotiations.

It is important that the persons who are going to use the documents should have all their entitlement as soon as possible—if necessary, before the final execution of the contract. For the quantity surveyor this usually means supplying the architect and the contractor with the conditions, and the contractor with the priced bills as well. In addition, the quantity surveyor will have his own copies and, like the contractor, will hope to receive drawings and any specifications in the fullness of time. The client's copy will be the contract copy proper and here it is helpful, in private practice, for the architect and the surveyor to have the custody of the documents which they themselves prepared, so that they can show them and explain them to the parties on request. In the case of an official client the question of custody will usually be prescribed.

With all these dispositions made, the documentary scene is set and all the *dramatis personae* are ready to play to the full their new parts in the next act.

There are however some further collateral activities in the cost control field to carry through.

Re-aligning the cost control data

With the priced bills or other basis to hand, it is appropriate to reassess the cost plan and other control apparatus. If the bills are in a suitable elemental form, it may be a straight case of substituting them as the medium of control. If not, they may still have been so arranged that a subsidiary analysis throws up the main controls required. Otherwise the original cost plan can be repriced to reflect the price levels in the priced bills. Even if the two totals are close, the detailed distributions may differ enough to make this worthwhile.

How necessary this exercise may be will depend on the expected extent of variations — the more extensive they are likely to be, the more necessary does comprehensive control become. For relatively small changes, approximate estimating on the basis of variation orders may suffice. Variations apart, a reassessed cost plan provides better data for future cost plans.

A rather different exercise that some clients will require is a forecast of their expenditure for cash flow purposes. This calls not only for an analysis of the contract sum, but also for an assessment of variations, claims, nominations and fluctuations. If it is to be reasonably reliable, it also needs data from the contractor's construction programme, especially if the sequence of work is able to be varied fairly widely. Rates of expenditure on nominated work are most vulnerable to fluctuation. Extensions of time are best ignored: they delay the contract expenditure and it is better to reassess the forecast as and when they are granted. If the same causes lead to claims, the rate of expenditure may not drop much in any case — although the total will obviously increase. The appendices to Chapter 24 give simplified examples here.

PART 4

THE POST-CONTRACT STAGE

CHAPTER 15

Some general principles

With the acceptance of the tender for a project, a profound change comes over several relationships. Before this event the client has had contractual relationships with only the architect, any consultants and the quantity surveyor. If he chose to withdraw he stood to lose only fees incurred to that time and the results of work mainly carried out on paper. Under their conditions of engagement the persons whose services he no longer required would have charged just for the pre-contract stages of their work and for any work specially done in preparation for later stages. Unless their contract with the client was unusually stringent, they would have recovered nothing for loss of future business and would simply have had to grin and bear it. Even with the signing of the building contract, their position will remain the same in this respect.

The real change of relationships centres around the heretofore tenderer. So far he has been someone seeking for business, with no absolute assurance that he will get it and while this stage persists, he has no enforceable claim against anyone if his prospects fade away into the mists—not that this looseness should encourage clients or their advisers to make frivolous invitations to tender or negotiate. But even while the ink of signing is still wet upon the contract, or so soon as the letter of acceptance is in his possession, all has changed. He has become the contractor, a party to the contract with the client, who in the same moment has become the employer or, perhaps, the authority. In these new circumstances, withdrawal for the client is an expensive operation; even before work begins on site or in the yard the contractor will have a valid claim for a measure of loss of profit on the business foregone. When work actually proceeds there will also be the cost of what has been done and the cost of removing from site.

At this stage also the architect and the quantity surveyor don the caps which they wear under the terms of the building contract. Although they are not themselves parties to the contract, they have particular responsibilities to the parties under its machinery; these are additional to their own contracts with the client, and change their relationship with the contractor. Previously they were concerned primarily with seeing that their client secured a sound commercial arrangement, while observing a proper code of conduct towards the contractor; now they have a number of specific duties

147

to fulfil towards the contractor and in a number of ways have to stand quite impartially between the parties. While this is generally true under all standard contracts, the position is not always so clear under the GC/Works/1 Form where the superintending officer is manifestly a part of the client's organisation (see *BCPG*, Chapter 27). Not all contracts identify the quantity surveyor and none mention consultants; any such un-named persons must be read as functionaries of the architect and, in practice, with only such authority as he delegates to them. This is discussed further in Chapter 4.

Not only do relationships change, but also hard physical facts. A change of mind no longer means the simple rubbing out of a line or the insertion of a few words; it may now mean knocking down hard materials or the disorganisation of complex activities, all of which is expensive. Of this more will be said later.

What may be emphasised, as commencement of work is considered, is that no contractor can make the best start in terms of effectiveness and economy if he is given inadequate time to build up his team, plan his attack and gather his resources. While pre-contract haste may be worth while in some cases, there is nothing really to be gained by then hustling the contractor on to site. He may come to secure a token start before some financial year runs out, or before some statutory time limit expires, but he will then need time to get properly set up. All that he will have done for himself is to have lost part of his contract period. It is difficult to generalise, but less than three weeks from the acceptance (not the receipt) of a tender for even quite a modest job should be considered minimal. For a project requiring special staffing or plant or materials on long delivery considerably more may be justified. In some European countries there is a regular pause of two or three months between entering into the contract and starting on site. While this is partly to allow the contractor to carry out his side of detailing the architect's design (as is the practice there), it also allows a sensible build-up in other ways.

Relationships with the team

During currency of the project the quantity surveyor will have frequent dealings with various members of the team. Many of these dealings will be extensions of the activities that were going on in the earlier stages; cost control and other advice to the client and the architect are obvious examples. Often they will proceed quite informally or follow such of the usual patterns of communication by letter, drawing or word of mouth as may be appropriate to the subject and its significance.

Dealings that particularly involve the contractor are discussed in succeeding chapters. There are however times when it is appropriate, to aid progress and co-ordination, for a good group of persons to meet rather more formally. The result is the site meeting. There will often be two forms of this; a general meeting at which design and construction are both represented and a more specialised meeting at which one element or the other alone is present. The design meeting will normally have the architect as its natural chairman, while the contractor will equally clearly lead the meeting at which construction arrangements are under discussion. Where the two elements gather in one meeting there is something to be said for either the architect or the contractor taking the chair, since one or the other may have a better overall view of everyone's needs at a particular time. Whoever sits at the head of the table, it is important that justice be done to all; even the smallest sub-contractor should have a fair hearing. The resultant minutes should be fair and should also be clear and accurate.

The great burden of such meetings will be to resolve points of design and progress. While these matters may be of interest to the quantity surveyor, in alerting him to unrecorded variations and to potential claims that he will ultimately have to settle, to a great degree they will be to one side of his most immediate tasks. A majority of surveyors therefore tend to regard such meetings as an exercise to be avoided if possible. This may often be a reasonable attitude, although there are other considerations. It may seem out of place here to consider the character of the client, but there are some who if they attend a meeting will expect to see all their advisers present and looking to their duties. Again, if the quantity surveyor aspires to greater eminence in the team, then perhaps he must be prepared to 'do a Parkinson'. Much will depend on the type of project and the degree of detailed involvement of the client in its day to day progress; some industrial clients are closely concerned and need everyone immediately to hand.

Then there are those jobs where things do not all run to the rule book and a keen contractor's surveyor may have considerable, and perhaps unfair, influence in a meeting. In such a case, the only way to redress the balance will be to 'double' one's opposite number. Even where no complications are expected from any direction, it is helpful to attend the first few meetings; this allows the others to get to know the surveyor and for general confidence to be engendered. Any routines can then be set up over daywork sheets and the like, and any immediately apparent oddities in the bills of quantities can be cleared up. Discussion of estimates for variations and measured rates should, however, be avoided in such mixed company: some present will have no interest or the wrong interest and seldom will all the data or time for the exercise be available.

Where the quantity surveyor is resident on site, or already visiting on the meeting day, he may be more inclined to attend since a special journey is not involved. Whether he attends or not, he should ensure that he is on the regular circulation list for the minutes and that he reads them when they arrive; they should be sufficiently detailed to warn him when he will need to note the next meeting in his diary. Where the programme is being disturbed, the site meetings are a key place to find out where everyone stands.

Relationships with the site

On a sufficiently large and distant job, the quantity surveyor's relationship will determine itself—he will be resident and see it only too frequently. Where there are extensive site works to remeasure, the acquaintance may become particularly close.

The great majority of contracts are not so demanding and some may hardly seem to warrant attention except at valuation times. For the smaller job this may be true, provided that the valuation visit is used to pick up variations that will be covered and otherwise lost, and that the foreman is allowed to voice any problems that he has. Even with the smallest of projects it is essential to pay special visits when ground works are under way. Where a job is of reasonable size, it will often need at least a fortnightly visit throughout most of its progress. Apart from contributing to settlement of the measured account, these visits help to build up a picture of progress that will be invaluable in assessing claims and, incidentally, lead to greater confidence at that stage, if it comes. Also, there are few surveyors who do not learn a great deal about building, and even about what they have taken off, by such forays.

Two persons on the site usually sustain a special relationship with the quantity surveyor. One of these is the clerk of works; whatever may be his strict position under the contract, he is of great help in filling in details for the settlement of the final account, quite apart from the signing of daywork sheets. It is important however not to rely on this gentleman to perform measurement duties which are the responsibility of the surveyor and, in passing, are covered by his fee. It is quite sensible to ask the clerk of works, if he is agreeable, to note the depth of an odd 'soft spot' with the foreman; it is quite another matter to delegate to him the recording of every drain depth on a major housing scheme. If the matter requires an interpretation of the bills of quantities or the Standard Method of Measurement, then it is doubly unfair to all concerned.

The other person is the contractor's own surveyor. Here the normal principle should be to deal with variations in his presence, unless he agrees otherwise or defaults in attending. Most groups of surveyors manage to arrive at a harmonious working arrangement, helped no doubt by their common background and outlook; this should therefore be a friendly, but reasonable restrained alliance. It is not desirable that either person should seek any type of ascendancy over the other that will create undue constraint at some crucial stage of settlement. This is not just a matter of lavish hospitality; it may arise as easily out of a difference of temperament or age and has to do with respect for the other as a person. It is particularly sad if an over-friendly time ends in a harsh settlement accompanied by a cry of 'et tu, Brute'. It is better to end with a rapture that a rupture.

Relationships with the budget

Once the basis for financial control has been revised as indicated in the last chapter, it is vital to work to it and from it throughout site progress and settlement. Many steps are obvious and may be deduced from later chapters. Only the main framework of control need be outlined, on the assumption of a fixed price contract. It need hardly be said that a contract based on such devices as approximate quantities or prime cost needs the most rigorous control.

The primary cause of financial change is likely to be design modification, with the direct effect of introducing variations. While estimation of the value of variations is fairly straightforward, those involving new materials or techniques and those likely to be dealt with by daywork are potential sources of over-expenditure. Even a variation that looks uncomplicated on paper should be examined carefully to see whether it involves work that has not been clearly expressed, but which is inevitable. Beyond this there is the possibility of having to alter work already executed or of introducing ripple effects by disturbing the even working of those engaged on adjoining or subsequent work. Further still there are the ever-recurring problems of orders issued late and causing 'invisible' expense, such as uneconomical ordering, and of orders not put in writing but carried out. While many of these are individually small and just irksome items in sweeping up for the final account, in sum they may amount to a substantial addition and also may have a combined effect on the contractor's programme so that a 'claim' situation occurs.

All PC sums and items must lead to instructions, although not necessarily to design variations. Almost inevitably they lead to financial change. A schedule of instructed acceptances is useful, along with a forecast of expected final amounts. Since there is a tendency for much work by specialists to involve their own design or other implications which the architect has kept to himself, it is a wise precaution to probe what is in the pipeline from time to time. While the quantity surveyor is strictly blameless if he waits to be told, he usually finds life easier if he can keep ahead of the game.

Progress or otherwise of the works can also introduce changes that affect the final account. Variations and other instructions and non-instructions can lead to the range of claims touched on in Chapter 20, even though not all matters leading to extension of time give rise to extra payment. Damage that is covered by insurance should prevent the client from being out of pocket, but if the client is to meet the expense and then recover the amount, there must be an allowance in the final account forecast. Where the client is acting as his own insurer the need is even more pressing, since he will be meeting the extra incurred. If the client wishes the contractor to advance the date for completion under one of the standard forms of contract, the contractor is under no obligation to agree to this. If he does and quotes a sum in consideration of changing a term of the contract, cost control is simple. If he quotes items of extra reimbursement, such as overtime, extra plant or floodlighting, all to be paid for as incurred, the quantity surveyor and the client may then find themselves on a slippery slope—and upwards at that! Clear definition of what may be done, and if possible how payment is to be made, should be established before matters get out of hand.

A special case where the programme is disrupted and extra expense incurred is that of determination of the contractor's employment. This is bound to cause additional expense which may or may not be recoverable from the contractor. Where there is insolvency, matters can assume run-away proportions as the case study in Chapter 25 illustrates. At this point any existing cost plan or the like may be consigned to the nearest waste paper bin. A fresh start is the only option.

The last major area of concern lies in the realms of finance and market conditions around the project. Cash flow as the client's concern has been mentioned in Chapter 14. It is not only a matter of whether the right funds are available to meet interim payments, but whether this flow represents the best rate of expenditure for the client in relation to the interest charges or other funding expenses involved. Once the contract has been placed on an agreed sum and time period, this is usually a matter about which little can be done. If it is desired that the rate of expenditure should, for instance, build up more slowly than might be expected and then attain a high plateau in the second half of the period, this must be stipulated in the contract originally. If the project consists of several direct contracts with some margin as to when they need be placed, the spreading of expenditure or its peaking may be possible if desired. Ironically many things that go wrong during progress tend to produce delay followed by attempts to regain the time list, and this means later flow of cash and some reduction of non-productive bank charges. However, on balance the client will not be impressed by this consideration.

Usually the pressure to gain early completion will predominate in the client's mind, but occasionally the balance of considerations may point in other directions. A client with changed market conditions for production or letting may even rather wait for completion patiently so that he can (if possible) collect liquidated damges as the more remunerative option. This may be unlikely, but is indicative of the thoughts which may be in a client's mind and which do not figure in a quantity surveyor's cost plan at all and seldom in his direct view of building costs.

One consideration that the quantity surveyor cannot overlook is that of inflation. In recent years it has thrown many cost forecasts into complete disarray. In view of the uncertainties inherent in high inflation periods, it is prudent to give forecasts on the basis of contract cost levels and the other items mentioned so far, and then to state separately the inflation allowance. As any part of this is used up in fluctuations or higher priced sub-contracts, an appropriate transfer from the uncertain to the more certain part of the forecast can be made.

As elsewhere, no attempt has been made here to describe the established techniques for operating cost control. Rather a number of policy approaches have been outlined that are relevant as the surveyor stands back from the project from time to time.This is in itself an important element in policy, both to ensure that something of consequence is not being missed and also to preserve a sense of perspective about what is going on. Otherwise cost forecasting can become distorted by an accumulation of small aberrations.

Two final points may be made. Cost control has been discussed as it affects the client's interests. This does not mean that those of the contractor should be ignored — indeed the very process of cost control should mean that the financial interests of the contractor over claims and the like must be reviewed carefully. Certainly cost control does not mean holding the contractor down to maintain the budget. The second point is that the best time for all concerned to know when the anticipated amount is veering from the budget is usually as early as possible — at least they have been warned and there is the possibility of taking action. Even any saving can be spent!

CHAPTER 16

Interim valuations and payments

The text of this book is related generally to the JCT contract, where appropriate. This fact is particularly relevant to the present chapter, and to those immediately following. Where other major standard forms differ from the JCT over the present subject, the key points of difference are summarised in the concluding part of this chapter. Clause references are to the 1980 edition of the JCT form, and the with quantities edition is assumed for discussion.

The case studies in Part 5 of the book illustrate a number of the principles discussed here and should be worked through in relation to this chapter and to the others in this part of the book.

Occurrence of valuations
THE MAIN SERIES
The JCT form takes little account of valuations in its formal scheme, unless formula fluctuations apply under clause 40. This clause introduces several amendments to the scheme and the added effect of these is taken up in the last paragraph under this heading. For the normal interim payments that the form otherwise permits, to finance the contractor during the progress of the work, it refers primarily to interim certificates, issued and apparently prepared by the architect. Interim valuations prepared by the quantity surveyor occur only when the architect considers them to be necessary. As common practice it is assumed that the architect considers valuations to be necessary, as they are on any reasonably sized job. In line with this philosophy the detailed scale of fees gives interim valuations as a separate, and therefore optional, ingredient.

Clause 30 of the JCT form is central to the question of certificates for payments, although several other clauses need to be consulted to obtain the full picture. It provides for two approaches to the intervals for certificates; the more common is that they shall be at regular time intervals up to and including the one after practical completion and here the most common interval is one calendar month, although any other agreed time may be laid down when the appendix is completed. For the great run of projects without repetition of numerous identical buildings, this is the most practical arrangement. The alternative system is to agree on stage payments—that is, to pay the contractor at defined intervals in the physical progress rather than at time intervals. Usually in this case not only the stages but the value payable at each will have been agreed on signing the contract.

This second method is very convenient for the small job not based on quantities; for an individual house it would be suitable to have three or four stages, such as first floor joists, roof coverings, and first fixing of services. For a large housing scheme of low rise units it is common to combine the two methods; while certificates are issued at, say, monthly intervals, each house is valued on a stage payment basis and usually only completed stages are included in a given certificate. This sometimes helps to encourage progress under the spur of securing payment for a whole parcel of work a month earlier. In such arrangements, however, the stages for individual houses are usually much smaller and may be as many as thirty for each house; this tends to reduce the spur, while making the cash flow more even for the contractor.

Where the usual monthly certificate system is being operated, it is a help to all concerned for valuations to be prepared on a fixed day of each month. It helps the quantity surveyor himself, although the considerations fixing the day may give him rather a hump of valuations at the same time in the month if he has several jobs running concurrently. Even this avoids having every week broken up by site visits. It is also a help to many clients to fit their payments to a particular pattern. Unfortunately their pattern may not suit that of the contractor; he is particularly concerned about making the most of his credit arrangements with his suppliers, who are usually allowing him to the end of the month following delivery. The contract provides for him to be paid by the employer within fourteen days of issue of the certificate. A valuation prepared in the first week of the month will include all materials delivered during the previous month, with possibly a few marginal exceptions. The resultant certificate should be issued during the second week and payment will be due a day or two before the contractor is obliged to pay the merchants. This is close timing; as valuations move, particularly into the second or third week of the month, so the contractor is having to find a larger sum to finance his operations.

If clause 40 brings formula fluctuations into the contract, the foregoing comments need modifying at several points. Firstly interim valuations to support all certificates become mandatory, in view of the dependence of the formula method on measured quantities. Secondly certificates and supporting valuations must be monthly, to allow fluctuations to be calculated properly. This therefore excludes the use of stage payments and also requires a regular date for valuations so that work executed may be related in even time periods to the issue of the monthly bulletins of index numbers. Otherwise work included early or late could attract the wrong level of fluctuations adjustment.

LATER VALUATIONS

The contract allows interim certificates to be issued after that following practical completion as and when they are needed, subject to minimum intervals of a month. It often comes about, in cases of remeasurement or severe variations, that it is some while after completion before preparation of the final account is far enough advanced to throw up additional balances. There will also be a certificate to clear the second part of the retention, or possibly more than one if partial possession or sectional completion has occurred. These matters of additional balances and retention may be duplicated in respect of nominated sub-contractors, who are also to be paid off finally ahead of the contractor under clause 30.7. Retention is considered further later in this chapter, while it and other aspects are illustrated in Case A in Chapter 24. For such reasons as these, the series of interim certificates is extended into the period following completion, even if then they are issued only at irregular intervals as required.

Preparation of valuations

THE GENERAL APPROACH

Individual surveyors vary considerably in their methods of actual working in preparing valuations but there are several points of common interest, however the detail is covered.

One is the part to be played by the contractor. He need not be consulted at all to satisfy the contract requirements, although he will have a right to arbitration if he is dissatisfied over a certificate in any matter of compliance with the contract. At the other end of the scale it is quite wrong for a surveyor, who is being paid to attend to the client's interests in not overpaying, to pass through a recommendation based upon an unchecked statement drawn up by the contractor. There are nevertheless good reasons for preparing the valuation jointly with the contractor or else asking him to draw up an initial detailed statement for checking. Some surveyors make a practice of writing a clause into the preliminaries of their bills to make this mandatory.

The advantage of bringing the contractor into the process is that he will know quickly what is the state of the various parts of the job. This is in the direct interest of the surveyor in saving time, but also serves the interests of the two parties by tending to avoid undervaluing. The client may otherwise receive an unwelcome shock when payments to catch up. When the two surveyors work together, it means that points of difference can be resolved at the time, although sometimes there is a tendency for work to take longer if one person or the other becomes obsessed with some minor detail. This may be avoided if the contractor does his work first and the quantity surveyor then checks the statement separately, since each takes his own time.

It also makes it easier for one or the other to delegate sections, such as materials on site, to an assistant and perhaps work quicker than his opposite number can. It may produce the situation in which the quantity surveyor has the 'last word' although any point of real issue should be raised with the contractor.

Where the two work successively, the contractor's statement may either be a little 'dated' by the time that the surveyor receives it, or may contain excessive allowances to cover anticipated progress during the time the surveyor is at work. In looking at this sort of figure, it should be borne in mind that the contract entitlement of the contractor is limited to work and materials up to seven days before the issue of the certificate proper. While it is sensible to include an allowance for later items, anything rash may clearly be deleted.

This attitude is rather linked with the general question of the accuracy to be expected in a valuation in any case. As already mentioned, unless there are formula fluctuations a valuation is optional and the inference is that something nearer to an enlightened guess is permissible. The whole system of payments on account is a modification of the basis of the contract as an entire contract at law. Only a reasonable accuracy is called for as the final reckoning is yet to come, except in the matter of formula fluctuations where the calculation is final, as further discussed in Chapter 19. This principle must be balanced against the twin duties of the surveyor to client and contractor. The latter relies on these contractual advances of his final entitlement to finance his operations on the way along, as has been suggested already.

The interests of the client need to be guarded so that there is adequate provision by way of retention against the possible default of the contractor, when every penny of the fund may be needed. For this reason there may be a temptation for the surveyor to keep his calculations low, even before retention is deducted; after all, he is likely to be blamed if there is an inquest. This is inequitable for the contractor and an unnecessary protection for the surveyor. Apart from the time lag before the issue of the certificate, there is also the permitted period of up to fourteen days for honouring. This should cover any minor aberrations, although not strictly excusing their occurrence. When there are formula fluctuations in the usual circumstances of rising prices, there may be a turn-about and the contractor may be tempted to curtail his cash flow if he can afford it and put work into a later valuation, thus securing any higher fluctuations reimbursement. These various conflicting pressures should help to keep the surveyor on a middle course.

A number of the elements entering into valuation calculations are considered below, the basic pattern being to work towards those figures which have to be shown in the architect's certificate for contractual reasons and which are included in the standard valuation forms available. As an absolute minimum these are:

(a) Value of work and materials, including preliminaries, formula fluctuations (if any) and nominated suppliers.

(b) Nominated sub-contractors, giving the amounts of each separately.

(c) Retention, showing that on (a) in total and that for *each* sub-contractor under (b) separately.

(d) Amounts *not* subject to retention, including traditional fluctuations (if any).

(e) Payments already made.

(f) Amount now payable.

In addition to these elements, it may be necessary to deal with value added tax on an interim basis. This needs no special mention here, while the general question of the tax is referred to at the end of Chapter 19.

PRELIMINARIES

There are two main problems relating to the interim valuation of preliminaries. One is that a number of items in that section of the bills represent costs which cannot be related to progress so immediately as it is usually assumed that measured items can be; the other is that contractors adopt widely differing methods in presenting the pricing of preliminaries. The one phenomenon flows from the other; some contractors show ten per cent or more of their work priced as preliminaries, while others show nothing at all and price their measured items that much higher. Even when preliminary items have been priced separately, the numbers of items actually priced, and the disposition of the money, may vary widely for the same bill total.

Valuation is at its simplest when the measured rates carry the whole of the pricing; payments will be a straight expression of quantity. When the preliminaries have been priced as a bill total only, or when just one or two items arbitrarily carry the whole value, either the pricing may be accepted as it stands or some analysis of the figures may be agreed with the contractor for interim purposes. When a fairly detailed pricing has been presented—which in this context may still mean only some half a dozen major items proportionately priced—then the figures must be taken as a working basis without question. If any analysis of a single total is contemplated for this purpose it is best to agree it at the tender examination stage to avoid later disagreement. It is also best to keep this analysis for interim purposes distinct from any amendment of the prices to be shown in the contract bills, since these may be used for calculating adjustments of preliminaries in the final account, where greater precision and even different principles may be required.

Once a workable analysis of the preliminary bill is available, the surveyor is left with the exercise of converting the various sums into payments on account. Broadly, the items represent costs which occur in one or more of the following three ways:

(a)   As a lump sum at a defined time in the contract period.

(b)   As a continuing cost spread throughout progress and related to time as a straight or curved graph.

(c)   As a cost incurred in a fairly direct relation to the expenditure on measured items, perhaps also influenced by nominated work.

If the only figure available is still the bill total, the choice lies between (b) and (c) or some blend of them. The simplest approach is to relate payment directly to measured value, leaving out nominated work, by a percentage at the end of the valuation. This safeguards the client's interest, since he does not see the benefit of the contractor's expenditure in the early stages of the work when measured value is usually accruing at a slower rate than in the middle time period. On the other hand the contractor may well contend for a time basis as in (b), since his expense in setting up the site is relatively high. Either approach has its strong points and can be justified. If the payments by either system are discounted to the end of the contract period there may well be little between the two financially. The real tension lies between risk to the client and outlay for the contractor in the early stages. If the contractor has chosen to price in this way, he has really forfeited his right to insist on any particular formula for interim payment. Any suggestion that an inordinately large share of preliminaries should be paid early is therefore quite out of place.

Where pricing is more detailed, the opportunity to philosophise is increased, but may produce significant differences only on large contracts. An example is given in Case A in Chapter 24. Items like 'Person in charge' are straight time-based items. Even here the sum inserted may be so large as obviously to include other site staff and it may be reasonable to allow for a progressive build up. Items like 'Holidays with pay' relate directly to labour and may be expressed as a percentage of measured work. Incidental related plant items such as 'Transport of workpeople and materials' lend themselves to the same treatment. An item like 'Clearing site at completion' is a once and for all matter and plain to behold.

Some other commonly priced items are more complex. This is the case with temporary works such as site huts, hoardings and temporary services. These are not related to measured items; they are however set up early and incur considerable expense early. Once set up they need maintenance, have running expenses, and represent capital tied up or hire charges going out. At completion, or before, they have to be taken away. In these cases, an analysis of each item may be justified where it is of any consequence.

Very similar to these items are major plant items like cranes and scaffolding. Where they are widely related to the physical work there is no point in treating them any differently. Sometimes they may be required particularly for one element of the construction and they may be allowed at a particular period of the contract accordingly. Any relation of them to quantities may destroy the point of pricing them as preliminary items and will usually be an over-refinement.

In the case of temporary works and plant, the employer has the right under the contract to use the equipment free of charge if the contractor's employment is determined (see *BCPG* Chapter 7) and it is therefore usually safe and reasonable to pay for setting up before the benefit is evident in finished work. The exception occurs when equipment is hired; here the hirers have a right to remove it and not necessarily to accept any assignment of the hire agreement. Payment of a heavy setting-up allowance under an early valuation may thus lead to later embarrassment and should be avoided. The ownership of any major items should be checked.

Another set of items is related to insurances and payments for water and other temporary supplies. Here again the contractor may have to pay out the whole or the large part of his expenditure very early. Whether to pay for these items as a lump sum or progressively will again be determined firstly by whether the benefit will pass to the employer at determination. Insurances will lapse, while sums for water will carry over. Where payment is related to metering, progressive payment is obviously appropriate. These principles are equally applicable where the obligations have been priced as percentage additions on the bill summary.

Where the works fall behind programme the payment of preliminaries, in so far as it is related to time, should be set back by the same interval rather than be spread over a longer period more thinly. This avoids paying too early. It is also a fair reflection of what happens; if the delay leads to a loss and expense claim the contractor will receive an additional sum covering the prolongation, if it does not he will have to bear the cost of the delay in all respects.

MEASURED AND OTHER WORK BY THE CONTRACTOR

The detail of bills of quantities provides all the information that the surveyor has or can reasonably need to value the physical work for interim purposes although the contract does not specifically require valuation to proceed on this basis, except where formula fluctuations are being applied. It can only be assumed that the items are priced realistically and reflect the actual cost of execution and the value of the works to date, to client and contractor alike, if either were to default or to determine under the contract. This assumption may fairly be modified where work under a bill item represents only part of an operation, such as excavating without refilling, and only that part has been done; usually 'value' will be taken as covering what has been done but in special cases payment may be held or reduced until the risk stage has been passed.

The wider implications of this assumption are concerned with the philosophy of measurement and pricing and have been alluded to at the opening of Chapter 7. Thus, for instance, in a tall building work on the lower floors will cost less per unit quantity to produce than work on upper floors, but the work will usually have been measured together and priced at the same rate. Again in a building with wings having variously large and small rooms the unit cost will vary, but the work may not have been separated in the bills. Even when the costs of hoisting and other distribution have been realistically priced in the preliminaries, differences due to labour will still exist. The contractor is on fairly strong ground in arguing for no distinction in valuations if separation was not considered worthwhile for tendering purposes, but a distinction could be made especially if the preliminaries have not been priced realistically.

A narrower issue is that some sections of the bills may have been deliberately priced high or low; this has been discussed in Chapter 11, under 'The structure of pricing'. If the appropriate action was not taken when examining the tender there is little that can be done at this stage of the proceedings, beyond watching particularly closely the dangers of allowing excess quantities.

Where firm quantities are available, and stage payments do not apply, it is usually possible to use these as the basis for calculation by apportionment. Since a valuation has a relative accuracy only, this can often be done using percentages assessed by inspection of the stage that the work has reached. Each work section or the like will be taken separately or, where it contains diverse types of work as in the case of finishings, then each subdivision will be considered. In the early and late stages of the work covered by any such section it may be that the work done or outstanding will be piecemeal and can best be assessed as an entity without using the apportionment method. If it is outstanding work that is in question, the resulting value will then be deducted from the bill total to arrive at the valuation. The order of the original bill should be followed as a general rule; this will lead to easier comparison of one valuation with the next. Indeed, it is convenient to bring forward section totals from one valuation to the next as sections are physically complete or nearly so. This practice should be treated with care, since it is possible by this means to introduce cumulative errors if what was included in a total is not checked when it is adjusted in a later valuation. There may also be problems in checking back through a whole series of past valuations to see exactly what is included in the present case. It is therefore desirable to prepare the valuation in full at intervals, to dispose of too much dependence on the past.

Variations should be taken into account by suitable adjustments, since these are to be so allowed under the contract (see *BCPG* Chapter 9). Where a project has sections subject to remeasurement, or is entirely so, it will be necessary to rely on the running work for the final account, brought up to date by approximate estimates of sections not yet measured, or measured in advance. Daywork and fluctuations not on a formula basis should be dealt with similarly.

In a contract in which formula fluctuations apply, the whole question of accuracy needs more stringent examination as has been indicated earlier. In addition the need to analyse the valuations into 'work categories' under the formula rules means that bulk totals taken from the bills may need splitting and that there is greater disadvantage in lost clarity in bringing forward totals from earlier valuations. The more the bills stray from a trade towards an elemental layout, the more this is true.

MATERIALS ON SITE

Clause 30.2 of the JCT form requires the inclusion of unfixed materials in interim certificates and their valuation thus falls to the quantity surveyor. The practical problem that arises is simply that of finding all the items; some will be in sheds, some stacked in the open, some within the building. Here the assistance of the contractor's foreman or the clerk of works will be invaluable, as they have day-to-day dealings with the materials and can locate them readily. It may be helpful for one of them to draw up a schedule of materials, provided that the surveyor still checks the items. Again accuracy is relative, broad totals often suffice and some use of invoiced quantities with an allowance for materials already fixed may be made.

Materials may be priced at invoiced prices, unless traditional fluctuations adjustment applies to them, when the basic prices should be used. Any difference will then go into the fluctuations figure in the valuation and not be subject to retention. Where formula fluctuations apply, the use of prices for materials valuations other than the basic prices notionally contained in the contract base month index figures is in order, since in this case the fluctuations adjustment applying when the materials are fixed is subject to retention. A snag occasionally met is that the measured item incorporating the material may have been so underpriced that to allow the invoice price in a valuation will mean an overvaluation at that stage. Here the invoice should not be used, but the price should be related to the measured item. Only on a major item is this finesse really worth while.

Materials that are not properly protected may deteriorate or be stolen and these should not be included in valuations, even before any deterioration; this principle arises again in relation to defective work, and is discussed further later in this chapter. The contract also allows for the exclusion of materials brought on to site too early in relation to the programme, so that the client does not have to finance matters ahead of need. In both these cases the decision on what to allow rests with the architect, as standing between the parties in all such matters. In any case of doubt he should be approached before including marginal items or he should be advised when the valuation is presented as to what has been allowed and the value of this.

A special position comes about where the contractor has quantities of materials stored off-site, in his own yard or perhaps that of a sub-contractor. This may be due to cramped conditions on site or to extensive prefabrication, which is really a particular case of lack of storage room. The architect has an optional right to certify the value of such materials, provided they are ready for incorporation in the works and provided the stipulations of clause 30.3 of the contract have been met in most parts of the United Kingdom or the corresponding procedure has been followed in Scotland. However, unless the contract expressly gives the contractor a right to such payment, as may well be done in an industrialised scheme, the matter is entirely within the architect's discretion. In no case can the surveyor include such items without the clear direction of the architect (who must ensure that the quite detailed requirements of the clause have been met), even when he is otherwise willing to do so. A further question that may be of importance if a contractor is running near to insolvency is that of the doctrine of reputed ownership (see *BCPG* Chapter 11); this may well lead to caution in making such payments, although this will need to be balanced against the risk of precipitating the insolvency if the payment is not made.

Once the architect has agreed to pay for materials off-site they will be valued like any others, making allowances for the cost of transport yet to be incurred. Often there will be difficulty of access and here is a case where it is quite reasonable to ask the architect or his inspector to give a schedule and thus save a visit. One of them will have had to have inspected the goods to see that they were complete and in so doing must have listed the items he was prepared to sanction.

Strictly it is only the materials that have been paid for that pass into the ownership of the client, and even here the question is complicated by how good a title the contractor has acquired and can pass on. Several recent legal decisions have thrown this area of law into disarray (see *BCPG*, Table of Cases) and opinions differ as to the precise position. In practice the surveyor has little option but to operate the provisions of his client's contract, which takes no account of this particular quirk, and include all materials properly on site. As a materials are incorporated in the works, they pass into the ownership of the client in any case. If the question of ownership is likely to assume a greater importance because insolvency of the contractor or of a sub-contractor is threatening, then it may help to identify precise batches of materials in the valuation and get the agreement of the contractor to what is so listed.

NOMINATED SUB-CONTRACTORS

The normal calculation of sub-contract amounts will follow that of the corresponding items for the main contract. While materials on site will be included, they will not, in most cases, be shown separately on the main valuation statement; it is the overall sum that concerns the contractor. It is then necessary to advise the contractor of the sums that he is due to pay. This is done by listing the gross totals of payments to nominated sub-contractors and the amounts of retention in the valuation and then in the certificate. As a result the architect and client also learn the position on each firm, which may also be useful to them. The architect is then required under the contract to notify sub-contractors of the sums payable. Alone out of the many who may be employed on the site, nominated sub-contractors have a special limited protection in that the architect must require proof of payment of previous amounts and, if the contractor has defaulted, the client may pay the firm direct as considered below. If the employer/sub-contractor agreement is in use for a particular firm, direct payment then becomes obligatory. This measure of protection is balanced by the fact that the sub-contractor is not entitled to any payment without an architect's certificate to the contractor. Not only so, but the contractor should not make any payment without a certificate. The architect may have good reasons for withholding monies. There are occasions when the contractor has a justification for withholding some payment from a sub-contractor, although such a set-off is carefully hedged about in the sub-contract. If it is justified, the amount should be ignored by the quantity surveyor in all calculations and procedures about direct payments.

Although the duty to check on previous payments rests with the architect, most surveyors will do it as a matter of course when preparing a valuation. Where the contractor has defaulted in a previous payment, no reduction in the amount of valuation must be made. This is solely the operation of the client, who has to deduct the amount of default from the amount stated as due in the certificate proper, provided that he also pays the amount directly to the sub-contractor. The architect is however obliged to certify any such default so that the client may know that his option or duty exists, and the surveyor must therefore accompany his valuation with a statement of the default. The obligation to pay nominated sub-contractors direct at the time of a particular interim certificate is limited by the total amount otherwise due to the contractor and thus available, so that the client is not due to pay out immediately to an extent that puts him out of pocket. If there is more than one sub-contractor in the queue, a proportionate payment system is normally required by the contract and the quantity surveyor is involved in calculating amounts. This is illustrated in Case B in Chapter 25.

A number of legal issues arise out of these matters of direct payment, and out of nomination in general, under clauses 30 and 35 (see *BCPG*, Chapters 11 and 16). These warrant careful consideration when a contract sails into troubled waters. What the client is not in any case entitled to do under the contract is to deduct from a certificate the amounts of payments in favour of a sub-contractor which are included for the first time. Only when default occurs does a right to deduction exist and it is limited to amounts previously included but not paid. Equally, there is no right of direct payment as such without a default, unless the parties specially agree.

RETENTION
Clauses 30.4 and 30.5 cover the rules about retention. They are related to what will usually be a single percentage held back from the gross value of work and materials. Items not subject to retention are considered under the next heading.

There has been a progressive reduction in the commonly accepted levels of retention over the years, partly because of the progressive increase in interest rates and partly because the contingencies of default against which retention is held have been seen not to warrant such high levels. In the case of a real catastrophe the retention will soon be swallowed up anyway, although every penny does help. As a result of the level being lower (the current recommendation by the JCT is either 5 per cent or 3 per cent according to the contract value), it applies right through without any maximum level of retention being specified. It is possible to set a percentage to be held until a particular sum is in hand and then to hold no further amounts. This can be done if a special case justifies the finesse, and is illustrated in the case studies in Chapters 24 and 25. If it is done it will need some care in definition to allow for the second of the two ways in which the total retention is divided.

The first division is according to the stage that work has reached. The full percentage is to be held on work prior to practical completion, and only half that percentage on work between practical completion and the making good of defects. Thereafter no retention is to be held. This is simple enough if all work changes stage at once. If however there is phased completion there will be work in two or more stages at the same time, and possibly that in one stage will not all be due to pass to the next at the same time in the future. A variation on this theme occurs if any nominated sub-contractor is entitled to receive early final payment under Clause 35.17, in which case the level of retention changes from full to none in one jump if needs be. This instance is related to the stage reached by the sub-contract works and this is likely to be completely out of step with the other reductions. These various instances are illustrated in Chapter 24.

The other way in which retention is divided is between that held on the work and materials in general of the contractor and on that in particular of the nominated sub-contractors. This is necessary because the client has a fiduciary interest in the retention to hold its various components in trust for the separate sub-contractors concerned. It is therefore necessary also to back up each interim certificate with a statement breaking down the total retention. The architect then has to issue this statement to the client and to the contractor and, according to the clause, to each sub-contractor. This last provision taken literally means that each sub-contractor receives details of everyone else's retention, which is more than he needs to know or should know. It is better for the architect to prepare separate sub-statements for each sub-contractor from the quantity surveyor's statement. Again Chapter 24 is illustrative here.

In all cases of a reduction in the level of retention, the surveyor must ensure that he has adequate authority, since the change may not be obvious to others from his valuation figures and so highlight what he has done for adoption or otherwise. If therefore he does not possess notification already of the appropriate certificate of practical completion or making good defects, he should confirm with his valuation any change he has incorporated on the basis of any certificate of this type that the architect should issue alongside the interim certificate for payment. This is particularly important in the case of nominated sub-contractors, for whom the detailed procedures are slightly different and whose subsidiary position could lead to confusion.

It may be that the client exercises his right in particular circumstances to make deductions either from the retention held or from that being released by a change in percentage. As with deductions that the client may make from the net amounts certified, these deductions should be ignored in all calculations for valuations and certificates, since the client makes the deductions as his own adjustments to these figures.

ITEMS NOT SUBJECT TO RETENTION
It is the philosophy of the JCT form to provide under clause 30.2 that payments adjusting the contract sum, but not resulting in a change in the physical work, shall not be subject to retention. The main cases in question

are payments for loss and expense that the contractor incurs due to disturbance, and payments for fluctuations calculated on the traditional basis
only. This last is notionally double-edged, since deductions from valuations
for fluctuations do not reduce the amount of retention held.

Among the other items of additional payment to be included gross are
statutory fees and charges levied on the works through the contractor, the
cost of opening up work and testing where the contractor is not in default,
and insurance premiums paid by the contractor where the client fails to
insure. All of them arise out of matters that are not the fault of the contractor
to say the least.

Particular considerations

There are times when the surveyor may be in doubt as to whether to make
certain deductions from his valuation totals. One of these, the case arising
where the contractor has failed to pay a nominated sub-contractor, has
already been discussed in this chapter. It is possible at least that the
contractor has a legal, if not a moral, right to take the drastic action of
determination under a hard reading of the contract if the employer exercises
his option to make such a deduction from the actual certificate presented to
him. This would be particularly unfair if it were the case, and calls for an
amendment of the contract if it is ever held to be so. The next case below is
subject to the same uncertainty.

LIQUIDATED DAMAGES
(see *BCPG*, Chapter 6)
When the contractor overruns his completion date on the contract without
securing an extension of time, he becomes liable to pay the amount of liquidated damages provided in the contract. This is usually dealt with by deduction
from monies due to him as they arise under interim certificates. It is not
within the authority of the quantity surveyor to decide that the contractor is
liable; this depends upon the issue of a certificate by the architect under
clause 24, stating that completion has not been achieved. Even then the
quantity surveyor must not deduct the sums that accrue, nor even may the
architect. It rests entirely in the discretion of the client as to when and how he
recovers the monies, if at all.

The prudent procedure therefore is for the surveyor to advise his valuation
without the deduction, but to mention the damages to date in a covering
letter. The architect should then adopt a similar procedure in issuing his
certificate.

When there has been sectional completion or partial possession, there will
be a reduction of the level of damages that is proportionate to the value of
the part taken over or at least only marginally different in proportion.

However, any early completion by the contractor cannot be put forward as offsetting the extent of delay on the remaining sections: the benefit he derives from early completion is limited to early reduction of retention. The late sections must continue to attract a liability to damages strictly in their own right.

DEFECTIVE WORK AND MATERIALS

A different position exists when work is found to be defective at the time of preparing an interim valuation. The value of this work should not find its way into the corresponding certificate, since the architect has to include only the value of work 'properly executed'. Since a certificate does not have a final significance under clause 30.10 when it is interim, this power of exclusion will extend to work which was included in a previous certificate and has now been found to be defective. One of the purposes of retention is to allow for just such discoveries to be adjusted without overpayment.

The decision over quality is always with the architect and thus beyond the surveyor, who may adopt one of two courses. One is to include all work carried out but not condemned at the time and to advise the architect that this is his standard procedure. If so he should reasonably comment on any work that strikes him as suspect and give its contract value in an accompanying statement. The architect in his certificate may then allow a deduction for this work if he intends to condemn it. Wherever it is convenient to do so, it is more straightforward to ask the architect before preparing the valuation and so exclude the work in the first place. This saves having to go back through records to adjust them in the light of the architect's decision and may also be less irritating to him. It still avoids misunderstanding if the valuation is accompanied by a statement confirming the action taken.

The same position arises when materials on site are possibly defective or otherwise in doubt, as has been noted earlier in this chapter.

Differences between contract conditions

Much of what has already been written will apply equally when dealing with contracts not based on the JCT form. The other leading forms do however differ in a number of important points of detail.

FORM GC/Works/1
(see *BCPG*, Chapter 27)

Conditions 40 to 42 of this form cover interim payments, among other matters. The task of preparing valuations falls upon the contractor specifically, although the superintending officer is to check them. The normal frequency is written into the conditions as 'not less than one month', but where the contract sum exceeds £100,000 an intermediate advance may be made, based on an estimate only. This will be prepared unilaterally by the superintending officer and is easily assessed as a conservative proportion of the preceding detailed valuation.

The contents of 'advances on account', as the conditions term them, are essentially the same as in the JCT form, although extra expense amounts may be included only when they have been agreed. The main differences centre around the calculation of the reserve, as the retention fund is called. The conditions give a fixed level of 3 per cent for the reserve, to be calculated on the total value of work executed and other amounts. There is however a separate reserve at the rate of 10 per cent held on materials on site, no payment for materials off-site being allowed under the conditions. At completion half of the main reserve is released, while that on materials will have disappeared by then.

Although the calculation of the valuation as a whole is not affected by the point, it would also appear that the contractor is not empowered by the conditions to hold retention on nominated sub-contractors and possibly he could be penalised if he did. There is no power to pay any such firm direct if the contractor defaults over payments, although there exists a much more severe power to withhold further payments until the contractor clears the position. Since the contractor is not due to pay subcontractors until he has been paid himself, the opportunity to check whether he is paying on time is bound to be delayed in any case.

These conditions do not make the same distinctions as the JCT form over items that may be paid with or without the deduction of the reserve. Thus sums for additional expense and the like will rank for a deduction. The position over variation of price payments — that is fluctuations by another name — is rather untidy. As in the JCT pattern, these are subject to reserve when they are calculated by the formula method. When they are calculated by the traditional method the superintending officer may only pay them in the interim advances if the parties so agree. This is usually done, but then the reserve is not applied to them, as with the JCT amounts on this basis, but as of custom since the conditions are silent on the point.

While the first half of the reserve is released at completion and any final balances at settlement of the final account, the conditions are not clear as to when the second half of the reserve should be released and some peculiar interpretations are possible (see *BCPG* Chapter 27). However the authority (the term for the client) can make payments at will and this usually means in relation to the clearance of defects. But this does mean that the surveyor must obtain authority to allow for any releases.

Also at completion of the works, the final sum is to be estimated and any balance thrown up is to be paid to the contractor then. After this there is no provision for other advances on account, as distinct from releases of reserve, until final settlement. The estimate of the final sum thus calls for some care.

This form differs from the JCT form by including value added tax within the main stream of conditions. It will thus be allowed for in advances, but will not be subject to deduction for the reserve.

THE ICE CONDITIONS
(see *BCPG*, Chapter 28)

The ICE form allows monthly payments, with the contractor responsible for preparing a statement in detail for the engineer at the end of each calendar month. There is a proviso that the payments are not due unless the sum for the month is at least as high as a figure set down in the contract documents. Payments for extra expense and the like are to be included, and there are arrangements for paying for off-site materials. Retention accrues at 5 per cent initially but tapers off to 3 per cent as the value increases. It is held on all the elements in the valuation, half being released at completion and half after the maintenance period ends, whether defects have been cleared or not. If they have not, then a sum will still be held back to cover them. When there is sectional completion 1 ½ per cent is released on the part completed. The form also envisages the possibility of late completion of parts of the work, without this being the fault of the contractor; in this case also there may be a partial withholding of retention.

Under the form, both suppliers and sub-contractors are termed nominated sub-contractors and this overall class is then subjected to retention held by the contractor. There is a right for the engineer to pay nominated sub-contractors direct and make deductions if the contractor defaults over paying them (a different procedure from the JCT system, although with the same effect), and so the level of retention becomes of immediate interest to the engineer or the surveyor exercising the discretionary right of checking on the correctness of payments actually made.

Beyond these points the conditions leave a number of matters undefined. These include fluctuations payments which, it would seem, should be subject to retention. Vagueness also leaves it open to continue to issue interim certificates during the maintenance period, provided that the value justifies this.

A point in which the ICE form differs from all the rest is in its specific mention of interest if there is delay in honouring interim certificates. This provision is included instead of a right for the contractor to determine in these circumstances. While he may render a separate account for any interest, it would appear to be in order for any sum to be included in certified amounts.

The form also differs from the JCT form in that value added tax is included in the main contract interim payments and is not a separate account. As with the GC/Works/1 Form, retention will not be held.

CHAPTER 17

Variations and final accounts based on bills of firm quantities

In what follows, as in the previous chapter, the JCT form is the contract directly considered. So far as the present topics are concerned, the other main forms have little that produces any real distinction in overall procedure. Form GC/Works/1 indeed is very close in effect, while the ICE form is far less precise than either of the others and thus permits the same approach, even if it does not demand it. The main point of consequence in the ICE form is the much more open power of the engineer to agree new rates where he considers the contract rates unsuitable. A reading of the clauses in these other two forms will therefore be sufficient to relate them to what is said here in relation to the JCT form in particular (see *BCPG*, Chapters 27 and 28).

There are other matters besides variations that enter into the preparation of final accounts, and several of these are looked at in the next three chapters. They include remeasurement, approximate quantities and schedules of rates; final accounts for prime cost contracts; all-in service contracts; PC sums and accounts; fluctuations; daywork; and payments for unexpected loss and expense.

The authority for variations
THE CONTRACT PROVISIONS
Variations are simply one of the matters over which the architect may give instructions (see *BCPG*, Chapter 4). His general power is given by clause 4.1 and the quantity surveyor may deal with variations only where the architect has acted under these powers or where the contract gives circumstances in which variations are deemed to have arisen. This is in accordance with the principle of the contract, which is made express by clause 14, that the contract sum may be adjusted only where the conditions so provide. That is to say the contract is an 'entire' contract in basic legal form, modified by its provisions.

Clause 13.1 gives the strict definition of variations and in so doing produces two categories. One is the introduction of changes in the physical characteristics of the project, that is broadly matters of design and specification. Most of these come by direct instructions of the architect, but others arise under clauses which define various events as though they were instructions, including complying with statutory requirements and correction of the contract bills

172

(see *BCPG*, Chapter 4). These variations are all quite clearly delineated in extent and can usually be measured, if not priced, equally precisely.

The other category of variation is the change of obligations or restrictions imposed by the client. These are limited in two ways: they must have been imposed in the contract bills and they relate to a tight list covering access, use of the site in particular ways, working hours and order of working. Variations of this type are likely to have more diffuse effects that cannot be reduced to measurement and that require special valuation. In the case of all variations it is the intention of the contract that any disturbance effect due to their introduction shall be dealt with separately as a claim under the procedures of clause 26, and this aspect is discussed in Chapter 20. The distinction between a variation and its effects becomes particularly difficult to maintain for this second category of variation and the main aim must be to cover the financial effects in the final account, subject to observing the relevant procedures for the clause chosen.

ERRORS IN THE CONTRACT BILLS

In particular, errors made by the contractor may not be adjusted in the final account, whether the priced bills were examined by the surveyor or not, when once the contract has been formed. This applies to arithmetical errors that may be apparent on the face of the bills, as well as to errors that underlie pricing and that are not so easily demonstrable. It also applies whether the error is an undercharge or an overcharge. It is however a commonly accepted principle that where work has been incorrectly priced then neither party can be obliged to agree that work in excess of the quantity in the bills should be carried out at the incorrect level. Where the rate is alleged to be incorrect, rather than there being a patent extension error, this principle will not be lightly invoked, but will be restricted to those cases where the error is of such magnitude that it cannot be doubted to exist.

The other source of error in the bills arises where there is an error in the description of an item or some matter of specification, or where a quantity is wrong or, more drastically, where an item is omitted entirely. Here the surveyor is allowed to treat the adjustment as a variation deemed to have been required by the architect and is thus spared the need to go to the architect and seek a variation order for the whole world to see. How he then absorbs the effect within the final account will be determined in part by the degree to which variations are itemised within the account or grouped together. Closely allied with the matter of error and sometimes inseparable from it, is the question of discrepancy between the bills and the other contract documents, or within the bills themselves. If the contractor finds such a discrepancy, he has to seek instructions from the architect under clause 2.3 and not make his own interpretation of what is to be done. Whether the item ends up as a discrepancy or an error, there is the possible consolation that it may not have been the quantity surveyor's fault in the first place!

UNCONFIRMED VARIATIONS

There are several types of item that may be presented by the contractor for inclusion in the final account where the surveyor is powerless to act without confirmation from the architect. One is an instruction given by the architect himself, but never put in writing; the procedure of clause 4.3 allows for confirmation at any time up to the issue of the final certificate and strictly speaking he should be approached by the contractor. Where expediency so dictates, the surveyor may well feel it quicker to make the approach himself, particuarly if the item has in any way been a sore point at some earlier stage and a little tact is needed. There will also be those cases where the architect may welcome guidance over whether the item is a genuine variation or whether it is already covered by some provision in the contract; while the issue of a redundant order may have no effect on the final settlement, it is at least pointless and may sometimes lead to a situation in which a payment is made inevitable. It may well be said that this should never happen; it is a simple fact of experience that it occasionally does.

Another case, at least as common as the last, is that of a direction given by the clerk of works. Under clause 12 such a thing is contractually of no effect, whether oral or written, unless confirmed by the architect within two working days. This is pushing things a little and means that often confirmation must come strictly under clause 13.2, not that it will look any the different for that. The contract recognises no variety of paperwork of the genus of site instruction or the like; these are however so useful for the minor items that the clerk of works can so conveniently clear as they crop up that there is a great deal to be said for the architect delegating a specific degree of authority to the clerk of works in this way. It is of course difficult to be specific here, in such a way that everyone knows without doubt that the clerk of works is acting within his authority and not exceeding it. Once he has given an instruction that has been acted upon, it is very difficult to explain to the contractor in particular that 'it should never have happened'. But unless there is such delegation, the quantity surveyor can do nothing in the final account. Even where the contractor himself confirms the clerk of works' direction back to the architect, the contract does not recognise this as conferring any degree of respectability on the act. Here again, the architect may well choose to lay it down that clerk of works' directions, spoken or in writing, are equivalent to his own oral instructions and thus eligible for confirmation by the contractor and with, of course, a right of cancellation by the architect. This gives perhaps the best of both worlds.

EXTRANEOUS CAUSES OF VARIATIONS

A variation may arise also from the discovery by the architect or the contractor of some discrepancy between the works as designed and statutory requirements, or from the action of the building inspector, district surveyor or other person with a right of control under statute or other external sanction. Here the contract, in clause 6.1, has procedures for the architect to issue instructions, either on his own initiative or in response to a notice from the contractor. These procedures are neither clear nor comprehensive and the contractor may act in good faith outside the (possible) letter of what is meant. In particular the clause is not entirely precise as to whether or not the contractor may act without a confirming instruction. While this situation usually presents no substantial cause of difference, the quantity surveyor should request architect's instructions on any doubtful item. In the case of compliance with a statutory requirement, there is the possibility of the architect securing a waiver and thus reducing the cost, or of a measure of re-designing that achieves the same object.

Beyond these persons, taken into account by the contract, there are other persons recognised as perhaps causing variations and whose acts the architect may choose to ratify under clause 13.1. One such may be the contractor himself, who may have acted quickly in some emergency where, for instance, groundworks were in danger or where an existing structure or its contents needed sudden care in some way that was not within the contractor's existing responsibility. The clause is not intended to cover the case where the contractor could not be bothered to ask or to wait, and simply went ahead with his own design solution to a problem. This is why the clause gives the architect an option as to whether to confirm or not. Another extraneous person causing such variations may be the client and here the architect may find it morally very difficult not to give a confirmation, however mortifying the item may be to him personally. Much will depend here on how well he has been able to educate the client and the contractor to accept that such things should never be done; a well behaved client will act through his architect in all situations. Obviously the two parties to the contract may agree to any such matter between themselves, but again the surveyor has no mandate to act until the item has been reduced to writing. This is even the case when the contractor simply leaves out some work and the surveyor notices it.

Strictly, architect's instructions covering the expenditure of provisional sums are not variation orders, since they may introduce no change in what was intended. Except where they result in nominated work, however, they will cause exactly the same processes to be set in motion and so they call for no separate comment. Often they will overlap with actual variations and the distinction will become quite meaningless in the account. Provisional quantities are a special case of a provisional sum and so, again, call for no further mention, although the procedure which is entailed when a complete remeasurement of a project is required is discussed in the next chapter.

Sources of variation information

From what has just been said, it is clear that the primary source of most variations should be found in the architect's formal orders. These will often be fairly brief and may well make reference to drawings accompanying them, and if there is some fresh material or process introduced there may be reference to a supplementary specification as well. Just as the original bills of quantities were produced from similar information, so should the variation account be. There is often a temptation to measure 'what the contractor has actually done' but the real criterion is 'what has he been required to do?' When the order and its supporting data are adequate, the surveyor should look no further. He may choose to check that the work on site is as the variation order and then measure it on site simply because it is more convenient to work this way; this is a different matter.

But there do come occasions when the written order does not clearly establish the extent or nature of what is required. This may be due to pure human failing or it may be one of those cases where it is necessary to give the contractor something approaching a blank cheque to solve some situation that can be foreseen but cannot be fully answered in advance. On any sizeable job, this sort of thing will usually be regulated by the supervision of the clerk of works, who should therefore be named in the order. In such cases, site measurement becomes inevitable unless adequate record drawings are prepared. It may be that the work can fairly be dealt with only on daywork and then it will certainly not be a case of working from the drawings, except in one respect; this will be to ensure that the daywork valuation, while covering the work concerned, does not overlap with the measured valuation or perhaps does the reverse by leaving a gap.

When the surveyor has got his sources clear, he will need to use them properly. Whenever possible he must rely on his direct observations; this is obvious when drawings are available, but is equally true when the finished product is the source. It should not become the established practice that the clerk of works becomes responsible for the measuring of all work that is to be covered up— this is as much the surveyor's job as anything else—but there are often circumstances of quick covering or of a distant site where it is only reasonable that signed sketches or notes from the client's man on the spot offer the answer. There may also be cases, although these should be rarer, when it will be necessary to take information from the contractor's staff in the absence of a clerk of works. Here the surveyor must use his discretion, based on his assessment of the individuals concerned. The problem may not lie in their honesty, but in their understanding of the type of information needed.

Where daywork is concerned, the quantity surveyor has no option; he is to work from sheets signed by the architect's representative. Should special circumstances arise which lead to the contractor being paid for the extra cost of working overtime, or for some other time-based element, then again this position will apply, although the contract does not mention it. At least in the case of daywork the quantity surveyor has a clear authority to decline to be bound by unsigned sheets, which may well be the fate of any which are submitted late. If it is, he may form his own assessment of the value of the work; if the work really is daywork this may still mean that he will refer to the sheets, but can use his discretion over times and quantities that appear excessive. The lateness of records does not debar the contractor from being paid at all, nor does the presence of properly signed sheets mean that the work may not be due to be measured or even be included in the original contract. Sometimes sheets are signed as being 'for record purposes only'; this is unnecessary because they always are.

Occasionally variation orders will arise that authorise the work to which they refer to be carried out on daywork. If once this happens there is little that can be done about it without some discussion. It is best to discourage the practice completely or, if this is not expedient, then to try to establish that there should be discussion over any order before it is issued, rather than afterwards. Some contractors become nervous over this issue, but it should not affect their contractual entitlement. Others may regard the practice as a suitable vehicle for the avoidance of losses on work which would be unprofitable if related to contract measured rates; here there is no entitlement to labour over.

Another case where the surveyor requires signed data occurs when fluctuations are being dealt with by the traditional method of totalling the hours of labour put into the works. Fluctuations are not a variation in the strict sense and are considered in Chapter 19.

Measuring and valuing variations
THE MEASURED ACCOUNT
Wherever the actual measuring takes place, there are certain points to be borne in mind; within limits these will apply to the subsequent valuing of the variations. It is a requirement of the contract that the contractor be given an opportunity to be present and, in apparent passivity, to take notes. Therefore, particularly when dealing with work that might soon be covered, an appointment should be sought in the first instance. Only if it becomes obvious that the contractor's surveyor does not intend to come to any

working arrangement should the quantity surveyor press on alone. If he does do this, then his records will stand in the event of a dispute except where there is a plain error. In the end, measuring together is to be preferred; it avoids later argument and may throw up some key element in the work that either of the surveyors might have missed if on his own. It sometimes takes longer, dependent upon personalities; here it may be quicker overall for one to measure and the other to check later. Where a contractor does not have his own surveyor he may prefer this system on that score.

In considering when to measure there is another element beyond the sheer convenience in arranging the surveyor's diary. This is the right time for measuring, which may be summed up as 'not too early, not too late'. When measurement takes place on site, things solve themselves; nothing can be done until work is erected in all its main outlines and then it must be measured before it is covered up — or demolished! When working from drawings the position is different. The main temptation is to put off work in the face of the current rush job in the office but occasionally there may be the chance to fill in a slack period by measuring off a number of as yet unbuilt variations on some contract. This can pay dividends, but has to be balanced against the possibility that the variation may be changed or even cancelled before it is implemented. Experience of one's architect or client may be the best guide here.

It is a fairly safe rule to say that the quantity surveyor should draft the final account, as in fact the contract requires. He is usually best equipped to relate it to his own bills and may need to present it in a particular form to satisfy the requriements of his client. If the surveyor drafts the account, he will then logically price it as well, as the contract infers. An account that is drawn to a total is the only one that should ever be sent to a contractor, to avoid the danger of someone altering a quantity, rare though this may be. At the pricing stage there is also a great deal to be said for putting one's own opinion of a contentious price forward first, rather than to start from someone else's.

Measuring should proceed by a series of additions and omissions, as the subject matter of the orders may dictate. Only when a section of work has been drastically altered should the complete omission be considered, although this is quite reasonable to avoid long-winded calculations. A straightforward remeasurement of what has been done will then follow. This principle is simply extended when it becomes expedient to remeasure the whole of the works, but here the discussion in the next chapter should be noted.

In presenting the account, it is usually convenient, and sometimes well nigh inevitable, that at least some variation orders should be grouped into one set of items. There will be some which are dealing with the same piece of work in such a way that the result is physically inseparable, such as when there is a 'variation on a variation'. Having taken these out of the reckoning, the surveyor will have many that may be billed quite separately. Some surveyors will do just this, and indeed may be obliged to by the wishes of the client, where he is an official one. Others will divide the account in the same way as the original bills; if these contained just individual buildings set out in work sections there may seem no point in doing more in the final reckoning. The original bills are a photograph of the project at a particular stage in its development. To know how it gets from this point to its final stage is the story given by the final account and may be said to be logically irrelevant if it is told with undue elaboration. Where logic rules, then nothing more complex need be considered.

One reason for sub-division may be the desire to bill and finally agree a section of the work ahead of the rest and thereby ease work at some later busy stage. This can be very helpful on occasions, even if it is agreed that some minor aspect of the section may need revision later. The other reason for division lies in the all too frequent need to hold an inquest on the job. If this is foreseen, the account may be divided into elements dictated by the grouping of cognate orders, all those for windows, for instance, being billed together. Those that have special significance in the inquest proceeding may be suitably highlighted, unless the reverse is indicated by the portents.

It is a truism that the contract says nothing about the presentation of the account, either in structure or detail. While the structures mentioned above are convenient, they are by no means essential. So far as detail is concerned, it is clear from the rules for pricing in the contract that measurement on the same rules as in the original bills must be used. Whether the results are presented in the same format is open; some surveyors favour direct pricing of the variation dimensions and their presentation as the final account. This is perhaps not so impressive to the average client and does lead to only one copy of the account being available, unless photocopying is employed. On the other hand it does save billing, even if the amount of pricing labour is increased. It also means that all information on the variation calculation is given in one place, even though the original draft bill of quantities does not do this unless a direct billing approach has been followed.

PRICING THE ACCOUNT

The contract rules for pricing are governed by the aim of keeping as close as possible to the original financial basis (see *BCPG*, Chapter 9). Thus measurement should always be used for additions in preference to daywork where the result is susceptible of reasonable pricing. The initial criteria for such pricing are whether the items are similar to those in the bills, both in character and in the conditions of execution and whether the total quantity of any item has not been significantly changed; if they apply then the original rates will also apply. If one or more of these criteria cannot be held to apply, then it may still be feasible to determine prices that are related to the bill prices on a pro-rata basis. Even where this is not satisfactory, it may be that the items are still of a common character whose prices are equally common and may be agreed between two persons without serious difficulty. This is termed 'fair rates and prices' by the contract, which would appear to mean a level of pricing that affords at least some profit margin but is otherwise in line with the pricing in the bills generally, whether this be relatively high or low. The term also embraces those items that may not be particularly common, but whose prices may reasonably be deduced from first principles.

In the case of omissions items, measurement is obviously applicable and usually the bill prices can be used as well. Only if again a significant change in quantity is produced will it be necessary to reprice the remaining quantity of an item on a pro-rata basis.

In each of these cases the main problem is usually found in the labour and plant elements of prices, especially where the conditions of working have been different from those of any items in the original bills that bear comparison. This is especially true where a variation is introduced out of sequence and leads to uneconomical working. Productivity can be seriously upset by lack of rhythm for the workmen and there may be a high concentration of supervision needed.

Alongside these considerations it needs to be remembered that other variations may introduce economies by helping working rhythm. A balance should therefore be struck between sympathy for every tale of woe and a hard, unbending adherence to the superficial letter of a weary contract. This same attitude is needed in relation to the remaining provision of the contract over measured valuation; this is that the variation of some items may mean that while other items are intrinsically unchanged, their cost is affected by a related variation. Substantially reduced quantities in what remains by way of other items of the same type may result, so that some major piece of plant is not being utilised efficiently. There may be the absence of work which was to share the use of a common scaffold. While these cases will arise, they may be counter-balanced by cases of the opposite significance. A pedantic adjustment of every such case may well cause much trouble that again can be avoided by a little 'give and take'.

When all has been done to see that measurement has its full play, there will inevitably be those items to which it does not sensibly apply. Where there are discrete parcels of work, daywork will have been recorded and the task of pricing is reduced to following out the terms embodied in the contract.

There are however those variations already mentioned (along with their disturbance effects) which change obligations and restrictions imposed on the contractor by the client, and these may produce wide-ranging effects on working conditions. For instance, there could be a change from one to two entrances for the site, or a reduction in the permitted working hours. It may be necessary to deal with valuation by what the rules term 'a fair valuation', the form of which will be dependent upon the nature of the variation. It is unlikely that a direct adjustment of contract prices will be suitable, although the calculation is likely to take account of work content, such as labour or transport. In some cases, preliminary items will be affected as well.

Any of these methods of valuation will take account of the direct cost of the variations only and there may be repercussions on other aspects of the contractor's costs. These are provided for under clause 26 of the contract and are discussed in Chapter 20, as being among the additional costs arising from several causes, the rest of which are different in nature from variations.

Some of these causes will give rise to an adjustment of prices for preliminary items, as is discussed. But there may also be a change in the volume of work in the contract, due to the cumulative effect of a series of variations, and this may lead to a reassessment of preliminaries. Where there has been an extension of time, the decision is fairly clear and it will be those items which have a time relationship that will be the targets for adjustment. Where there has been no extension, or it has only a limited bearing, it will be necessary to consider whether any of the value or quantity related items are significantly affected. Clearly the addition or omission of two square metres of brickwork or plaster will have no real effect on the site overheads. How great a change is needed to justify an adjustment is a matter on which surveyors will differ. A shift between contractor's direct work and that of sub-let trades or nominated firms may be more significant than the net difference in a particular case. While generalisations are apt to be rash, perhaps it can be said that the majority of surveyors will not consider an adjustment when the overall effect is less than five per cent of the contract sum. Since the pricing of preliminaries is quite the most variable aspect in bills of quantities, changes in measured quantities may in themselves produce a smaller or greater preliminaries adjustment. This factor alone is sufficient to ward off too dogmatic an approach to the question. The wording of clause 13.5 is silent on the point; its terms allow prices in preliminaries to be taken into account along with any others. In practice, the initiative for an increase more often comes from the contractor and it is probably fair to wait for this when dealing with a contractor who has a proper grasp of the structure of his pricing and its effects on his ultimate return. But where a contractor is obviously relying on the surveyor to guide him through to a settlement, then the surveyor should take the first step over an increase, as he will often need to do over any suggestion of a decrease. Points bearing on the actual adjustment are beyond the treatment given here, but the discussion in the preceding chapter over apportionment for valuations touches on relevant facets.

Some elements of the final account will inevitably appear in distinct sections of the document. These include nominated accounts and fluctuations, which are discussed further in Chapter 19. By virtue of their segregation in this way, these elements undermine the consistency of a division of the account by causes of variations, unless an excessive amount of sub-division is introduced into these parts or a subsequent analysis is produced. Additional costs that fall into a 'claims' category will often be calculated quite distinctly from the general measured account and may therefore easily be presented separately, as will be desirable from the point of view of any inquest on costs. If they are calculated by adjustment of rates in such a way as not to show separately on the face of the account, an analysis will be called for to throw up the required figures.

Finalising the account
AGREEMENT WITH THE CONTRACTOR

One of the peculiarities of the contract procedure is that it does not require agreement of the account to be reached at any stage. The architect is to state the figures derived from the final account in his own formal final certificate, after which the contractor has a period of fourteen days from receipt in which to give notice of arbitration if he does not agree. It is exceedingly doubtful whether anyone waits until this late stage before reaching an agreement, even if it may occasionally be subject to the resolution of some other matter than direct finance under the contract. Apart from the initial measuring and valuing, clause 30.6 gives two interim actions that are required to be taken, even though no agreement is to follow them of necessity. In practice it is expedient for the multiplicity of small actions that they embrace to go on throughout the duration of the work as well as after.

One of these actions is the supply of a copy of the statement of all the final valuations under clause 13 to the contractor, by the quantity surveyor who has done the measurement and valuation on which these bills depend. These bills appear to be the measured and daywork elements of the whole account and not to include the adjustment of PC sums and in theory possibly other items as well, which are listed in clause 30.6 as adjustments of the contract sum. The intention is to allow the contractor to mull the bills over before the final certificate descends on him unexplained. So far as the preparation of this part of the account goes, the responsibility for it and for any delay is squarely on the quantity surveyor, provided he has not been held up by the late production of invoices or the like by the contractor or by disputes over the ratification of variations by the architect. It is usual to provide in the contract that the period for the production of these bills shall be six months from practical completion, although this is only too often something of a butt for weak jokes on many jobs. It certainly needs great discipline on the part of all concerned to achieve.

The other action is the sending to the architect of all documents which the contractor has that affect settlement. These will include nominated accounts, fluctuations details (where traditional adjustment is operated), invoices supporting daywork and special measured rates, and sometimes insurance premium receipts. Not all of these will directly interest the architect, and he will usually be quite happy for them to go directly to the surveyor, who will need to deal with them all. In the case of nominated items in particular there may more often be points that the architect will wish to see and it is desirable for him to see all the accounts, whether they go to him from the contractor or whether they are passed on by the surveyor.

Whatever the devious paths by which the various components of the final account come together, it is desirable that they duly appear in a unified document and that the contractor signifies his agreement to the final figure achieved, either on the face of the document or by a suitable letter. This avoids misunderstanding over the final certificate. If there is some matter beyond the control of the surveyor which is still in doubt, such as a reference to arbitration or a defects liability problem or a claim, then the contractor should be invited to sign subject to the matter in question, whether it might lead to a straight addition to the account or to the reopening of some part of it. It is better to achieve as much agreement as possible as early as possible, so long as the effort is not likely to be entirely wasted. If the contractor is not prepared to sign the account or send a letter it is difficult to force the issue since he is not obliged to agree to anything other than the final certificate. Even if he does sign, it is open to question whether this has any binding force at law since it is always open to the client to disagree within fourteen days of the final certificate. For that matter the architect may not accept the account as drafted. However, most persons are prepared to let their word (or signature) be their bond or the quantity surveyor be their agent, as the case may be.

PRESENTING THE ACCOUNT

Much will depend on the individuals concerned as to what happens to the final account as eventually settled. As a minimum, the surveyor should expect to circulate copies of a statement to all parties, setting out the salient features of the finances in the settlement. At the other end of the scale, at least the parties to the contract are entitled to a full copy if they wish; this is reasonably implied by the proviso of clause 30.9 that the final certificate may be revised if arithmetical errors or the like are found. This proviso will be effective until the end of liability under the Limitation Act. So far as the contractor is concerned, he will perhaps wish to refer to the account for purposes connected with future work. At least, when copies are being prepared for others, the architect should be given one and there are strong reasons why he may want one in any case. It is always prudent for the surveyor to obtain authority for producing copies of the account, since their cost will be chargeable as an extra on his final fee statement and could be a relatively large item if the document is lengthy.

Once the account has been used to back up the final certificate, the work of the quantity surveyor under the contract has come to an end and he will be left to put away his papers as opportunity offers. These he must keep until the limitations period has run its six or twelve years before he is clear of his responsibilities to the parties. In particular he may be required to produce his details for audit, if his client is a corporate body entitled to require this.

The client, in the meanwhile, is obliged to pay the amount shown on the final certificate without waiting for audit to take place, if it applies. He has no power under the standard contract to hold any reserve and should recover any refund by the normal means when it has been established. At this stage Form GC/Works/1 is silent and the contractor may have little option but to wait. Under a JCT settlement, the client should be reminded that the account does not credit the amounts of any direct payments that he has made, or other contra charges such as liquidated damages. The final certificate arithmetic takes care of this in principle, but doubt should be avoided.

CHAPTER 18

Final accounts not based on
bills of firm quantities

In some instances it is appropriate to talk of variation when drawing up final accounts, but this is not always the case. The types looked at below fall into both categories between them; they are considered however in groups that reflect their degree of relation to firm quantities jobs in other ways. The pattern is thus similar to that employed in Chapter 12, dealing with the examination of the corresponding tenders.

Remeasurement, approximate quantities and schedules of rates

While three distinct bases are comprised here, they lead to what are very much the same activities in practice. Remeasurement has its own distinctive aspect, which can be disposed of first; all three then fall to be considered together.

By remeasurement is meant the result of a situation in which a firm bill contract is so changed that the only realistic course to follow is that of measuring the whole of the works afresh. This may occur where there is a fundamental replanning of work, and variations become a meaningless concept, but it may also happen where there are so many variations throughout the whole of the work that it is far more straightforward to start again. In this latter case it would still be possible to use the additions and omissions system, and this should not be lightly abandoned, simply for the convenience of the quantity surveyor.

Whatever the physical reason for seeking to remeasure, the permission of the client should first be sought on two grounds of principle. The lesser reason is that the fee payable to the quantity surveyor will be changed, possibly upwards, while the argument for remeasurement is that his work should be eased. (The fee position is discussed in Chapter 3.) If, however, variations would produce an unreasonable amount of intricate work, the request is sound. The greater reason is that the contract sum under a firm quantities contract is not to be altered except as the conditions of contract allow; while a remeasurement aims to produce the same answer at the end, it does mean that the contract sum is set aside and not adjusted. This change in the basis of the contract requires the agreement of both parties.

When once a firm bill is to be treated on a remeasurement basis it becomes no different from one labelled 'approximate'. The latter in turn is no different in how it is referred to from a schedule of rates. In each instance the work will be measured as executed, using the methods and items set out in the original document so far as possible and otherwise following the normal rules for valuing other work.

In measuring the whole, the surveyor must have the same authority for any item included as he has for a variation and he must use his sources in the same priority. Any suggestion that one simply and uncritically measures 'what has been put in' should be ignored. Where there may be a change of emphasis is on the question of when the measurement takes place. For the main structural and finishings sections of the work it is most economical of effort to wait until these parts are sufficiently near to completion for all to be decided, and for it to be too late for any changes to be introduced that would otherwise upset the measurements taken. They can then be measured also in the obvious overall sections so beloved of surveyors. Such delay may however mean that valuations are made more difficult each month, since an approximate measurement has to be carried out each time; there is always the danger of some serious error resulting from this process.

Two means of overcoming this problem may be used, although they both increase the final measuring labour. One is to go ahead with the measuring before the work on site, and to produce a bill that can be used as a basis for valuations. If there are any subsequent changes these are incorporated into the final account by an additions and omissions section appended to the main bill. This method is also a great help in tightening up the cost control system, always a thorny problem on such jobs, by giving a firmer forecast of how things are shaping. The second method is not so helpful here, although it is better than just an approximate estimate alone, and it leads to more work overall. It is to measure the work in stages as they are done, and to produce a final bill for each stage. Thus costs are known early, and only a comparative margin of work in progress needs to be assessed by approximation. Obviously, a much bulkier final account results; but the system has much to commend it on very large projects where the option of measuring further ahead may be precluded by the state of the design. The thickness of the account may be reduced in some cases by maintaining a running abstract.

In the case of drainage and services measurement, it is often possible to book dimensions in schedule form. Suitably arranged, these act as a source of running totals from month to month.

It can be seen that the form of the final account for complete measurement will be affected in its detail by the above considerations. Any presentation by sub-headings of orders will be out of the question, except for special sections. A grouping by trades, elements or work sections may be thrown up automatically by the measuring system, with the usual major structuring by buildings and the like. Any other groupings will be according to the requirements of the client or architect for accounting purposes.

Prime cost

According to which variety of prime cost contract is being used, so will vary the incentive for the contractor to economise that is built into it. These incentives have been discussed in Chapter 7; they are also bound up with the intrinsic efficiency of the contractor and with his integrity. All these things will be contributory to the final performance and the cost incurred to achieve it. Whichever pattern of contract is used, the processes of control and accountancy followed by the architect and the quantity surveyor will be the same, although the emphasis may vary according to the interplay of the other factors. Since the contractor is to be paid for what he expends, with only some reservations, rather than for what he produces, the architect is concerned directly with the work put into the hours and thus should exercise control over how the contractor organises his activities. This will usually mean watching that things do not decline into mediocrity, rather than taking over the reins directly. The JCT Fixed Fee Form (1976 revision) only suggests that this power is one of refusing to pay for over-provision and, even here, is rather gentle on the point (see *BCPG*, Chapter 24). In addition, the cost of remedying defects should be excluded.

The role of the surveyor is to deal with the accountancy side and thus to be concerned with the hours put into the work, rather than the work put into the hours. In practice, such a contract requires a close working together of architect and quantity surveyor, so that each may aid the other without transgressing into the other's field of responsibility.

In what follows therefore, a rigid distinction has not been made between the work of the two. Nor have the several variants of prime cost contract been distinguished; any real difference will lie in the need to adjust the fee, and this is simple enough, as shown by the appendix to Chapter 7. To arrive at the need for adjustment, there may be a measured account to be prepared, but this will be dealt with like any other measurement. Obviously, everyone will find things much easier if there is a resident clerk of works or a site checker; otherwise much will depend on the frequency of the visits made to the site. These should preferably be planned out for maximum coverage, while at the same time if they are somewhat unpredictable this may help in difficult cases of control and checking.

EXPENDITURE ON LABOUR

Labour is to be paid for on a basis which is usually related closely to the working rules and therefore the actual valuation of the hours is an essentially arithmetical operation, requiring signed wage or daywork sheets and a degree of patience worthy of a certain patriarch. So far as it is conducive to economy, the site should be manned by full gangs of men, who are not moved on and off the job in a manner best calculated to confuse the checking operation. Not only should these men be working industriously, but they should also be organised for efficient working. This is a matter not only of general supervision by the contractor, but also of the right grade of man doing the right grade of work so that, for example, tradesmen do not unload materials.

The cost of labour is also affected by the non-productive element of overtime, by bonus payments and by importation costs, all of which are at the contractor's sole option in a measured contract. There should be a close scrutiny of what is proposed for each item in dealing with prime cost. It is not a case of not allowing any such expenditure at all, but of keeping a balance between the total spent and other considerations, such as the practicable labour force and the programme. The JCT Fixed Fee Form provides for arbitration on these items, since they will have a bearing on the profitability of the fee. This point will be even more burning in the case of a target contract.

The central point at issue in all these cases is that the contractor does not have quite the same incentive to economy as is present in any contract in which he is paid for the result rather than the effort, as has been pointed out in Chapter 5. Some of the items are discussed in the next chapter, when dealing with fluctuations, where it will be seen that they have a different significance, since the contractor has his own incentive. Other items mentioned in that chapter occur in the present context with a similar significance; these are those related to work being carried out late in the programme or over-running it. In a prime cost contract, labour expenditure is paid for at the rates current when work is performed; thus later work will cost more. If the work over-runs the date for completion, the contractor should still receive a full reimbursement, on the grounds that he has either been granted an extension of time or that he is liable for liquidated damages. These latter should have been assessed at a level that takes account of the loss caused by the later completion, including the extra payment to the contractor due to his work being late, both during and after the contract period. Only where the contractor delays work during the contract period, without producing a commensurate delay in completion, will there be a strong case for rejecting the extra cost resulting from a wage increase or the like.

EXPENDITURE ON PLANT

Payment for plant will be made in accordance with a schedule of hire rates, such as that issued by the Royal Institution of Chartered Surveyors. The

JCT Fixed Fee Form defines what is meant by plant, but states that the hire rates are to be agreed. It is more than desirable that in a project of any size the rates too should be settled in advance. The schedule as used will usually provide hourly rates for plant in use, although it may sometimes be necessary to provide two rates, one for working hours and the other for standing hours. In general it is better to avoid the latter so that there is not any encouragement to leave plant standing on site after its use has really passed, or to use it spasmodically over a prolonged period. This is a particular problem with plant; labour will be paid off when it is no longer occupied, while plant that is not hired may be found to remain on site simply because it has to be somewhere. If the contractor has no other pressing job for plant to perform elsewhere, what happier home for it than a site where its depreciation will be covered.

Some items of plant are particularly awkward in this respect because, while they are major in character, they are sometimes used continuously and sometimes intermittently in frequent short bursts. Typical in this class are cranes and hoists, by virtue of their lifting function. To keep proper records of their working time is far too onerous; for these items it is better to agree a hire charge for the whole period that they are to be on site, as a lump sum or as a weekly amount. This has the advantage of absorbing the transport charges if desired and is the equivalent of the pricing of this type of item in the preliminaries of a bill, as is often done.

In view of what has been said, it will be clear that the closer it is possible to get to paying for all plant on a lump sum basis the easier will it be to exercise cost control. However, the circumstances that lead to the adoption of prime cost are often those that militate most strongly against such a solution, and so the best compromise must be found. Whatever the basis of payment, plant should be brought on to site only by agreement with the architect, and then retained only when its continued presence can be justified by outstanding work, unless the simple lump sum payment then applies.

EXPENDITURE ON MATERIALS

Materials are usually paid for on the basis of invoices and therefore at current prices. In the case of major items, quotations may be sought to regulate these prices, much as is done when a 'PC price' is included in a measured item. For the smaller quantity items it is sufficient to check that the invoiced prices are normal market prices. The cost of returning packings is admissible, unless otherwise defined, while a level of cash discount will be given. This will usually be for discounts 'obtainable' and thus does not entitle the contractor to receive an addition in lieu of discount where none is customary. Trade discounts should be deducted from the sums included in the final account, so that the client benefits by those. It is as well to make it clear in the original documents that annual and similar non-invoiced discounts will not be deducted, so that the contractor obtains the benefit and can therefore allow accordingly in his fee.

One or two further points may arise over invoiced prices. These should normally cover delivery, although this may be to the contractor's works if processing is involved. Unless there is some special storage problem on site, delivery to the contractor's yard for storage should be watched. In itself it is unexceptionable, unless materials are likely to suffer from multiple handling. But payment for transport from works or yard should be made only where circumstances justify it. It is also harder to exercise proper control over materials that pass in and out of a yard. Deliveries of materials in small quantities may arise in the closing stages of a job, but should normally be queried at other times, since such deliveries mean higher prices.

Materials coming from the contractor's premises can lead to other effects. They may have been in stock and thus not have an invoice. Here pricing at current prices is fair, but the contractor should pay the transport costs if the materials were chosen from this source as a matter of convenience. If materials have been processed at the contractor's works, then it must be checked that the payment for processing is not made under the headings of materials and also of works labour and plant.

The control of materials quantities should be carried out in one of two ways. There is direct physical control by checking deliveries to site and then checking that any materials removed for any reason are credited. This will be co-related with the invoices as they become available. In addition, it is usually feasible to carry out an overall check by measuring the work as executed and calculating the quantities reasonably needed. There is room for a fair margin in such an operation, but it will highlight any serious discrepancy in either direction.

Like labour, materials may be subject to increased costs and the comments about delays in the programme made for labour are as applicable here.

SUB-CONTRACTS

Nomination of sub-contractors produces no special problems in a prime cost contract and will be dealt with by following the procedural points covered in the next chapter. The nominated work may be related to any contract arrangement: the use of prime cost is by no means essential, so long as the sub-contractor himself is carrying out a defined parcel of work under defined conditions that suit the arrangement used. The associated financial matters are straightforward. Cash discount is usually allowable in the normal way. It may be necessary to add profit separately or, as in the case of the JCT form, it may already be included in the fee for the contract as a whole. Attendance consists partly of administrative and supervisory activities which will again be included in the fee element, and partly of site activities which will fall within the routine summation of labour and plant in the main.

Sub-letting of work is straightforward in that payment is simply made on the agreed basis for the sub-contractor concerned with any related allowance for the contractor, that is very much in financial terms as for a nominated sub-contractor. Under a contract like the JCT form this has the complication (as mentioned briefly in Chapter 7) that sub-letting may produce a different amount payable from that payable if the contractor had performed the work himself. This is because the agreed basis may be different, the JCT form for instance providing that 'the method of charge' is to be approved when the consent to sub-letting is given. When the work is specialised and the contractor is unlikely to do it himself, the question becomes academic. When however the work is within his reasonable competence, it will not be unreasonable for the agreed basis to be payment to the contractor at the same levels as for him performing the work himself, although the contract may well not require this, as the JCT form does not. If likely cases have been foreseen and agreed when the tender was under consideration, this otherwise post-contract question will be eliminated and possibly be settled on better terms.

Drawings and specification

For present purposes this heading embraces both the traditional small lump sum contract without quantities and the possibly much larger all-in service contract. They have in common the fact that the contractor has, as a general rule, carried the risk of errors in the quantities in tendering, since he will have prepared the quantities himself. In the examination of the tender this position is maintained, by excluding any quantities seen at that stage (and there may have been none) from any contractual significance. This has been considered in Chapter 12, along with the establishing of any schedule of rates for variations. It follows therefore that when variations are measured or otherwise ascertained, the extent of adjustment will be as from the drawings in the contract to the work as properly varied. This may be more or less than the deviation from the contractor's own quantities. Anything akin to remeasuring any part or the whole of the work, or otherwise reassessing it overall, should be considered only in the most dire of circumstances, since the contract sum will be set aside in so doing. What was said earlier in this chapter about obtaining the client's consent is all the more cogent here.

If a course of this nature is adopted, the original financial basis will have been so subverted that it will be in order to re-open the question of the accuracy of the original quantities if a schedule of rates has been derived from them. This may be done by agreeing correct quantities for the contract works with the contractor and setting against these the extensions in the original tender build-up. From this comparison may be derived fresh rates for the reassessment. This will have a result similar to that where there is no schedule of rates, but where there may or may not be some analysis of the lump sum into smaller sums.

There may, of course, in an essentially lump sum scheme, have been some section of work represented by provisional quantities or a sum. Here the normal rules will apply.

If the project happens to be an all-in service job, the division of design responsibility or its complete assumption by the contractor may lead to refinements or coarsenings of the foregoing procedures, as may the contract rules over discrepancies and divergencies (see for instance *BCPG*, Chapter 22 on the JCT form). This is especially true when the contractor assumes some unusual degree of responsibility for the cost of dealing with ground conditions. It also occurs when some part of the design is still to be developed by the contractor during the post-contract period. Here the dividing line between development and variation may be uncertain and lead to some latitude in settlement. On such matters it is difficult to set out principles and so the chapter must end, leaving readers to their speculations over the latitude involved.

CHAPTER 19

Special aspects of final accounts

There are two particular sides to final account work for the surveyor which have received some passing mention in the preceding chapters but have been left over to this chapter for more discussion because they are not properly variations and have aspects of their own. One is the adjustment of PC sums with the related settlement of nominated firms' accounts, while the other is the matter of fluctuations. A brief mention of daywork and value added tax is also needed. Beyond these are more ominous matters, considered in Chapters 20 to 22, which arise only when unexpected circumstances force themselves on the parties.

PC sums and accounts
EXPENDING SUMS

Nominated work may arise under a JCT contract in several ways (see *BCPG*, Chapters 16 and 19). There may be a PC sum for a supplier or a sub-contractor item in the bills of quantities and the architect can nominate against this; here the contractor knows clearly that a nomination is to be made and he will have priced for profit and any attendance accordingly. There may be a nomination arising out of an instruction over spending a provisional sum without, apparently, any restriction on the type of work that may be the object of nomination, so that here the architect has a power to nominate in respect of any suitable work that turns out to be required when the provisional element is defined. Lastly there may be a nomination introduced by way of a variation, and in this case the power is quite closely prescribed by definition of the categories of work stated. Effectively these are work of a similar type to existing nominated work for a sub-contract and the case of a sole supplier for a supply situation. Apart from the lack of an actual PC sum to deduct, and the need to agree profit and attendance items, this case causes no real difference in practice.

It would seem to be the legal position that the architect cannot introduce an entirely fresh nomination for extra work into a contract, that is to say, one not preceded by either a PC or a provisional sum, except in such a limited case as that of variation just mentioned. The contractor is otherwise apparently within his rights to insist that the work be treated as his own direct work, even though he may wish to sub-let it. Similarly, it is not within the architect's powers to issue a variation order omitting measured work and substituting the same work as nominated work, since this too is creating a new nomination. This must be distinguished from the case of omitting measured work and adding work of a type already covered by an existing PC sum. Only if the variation were carried to an extreme that could be shown to have disturbed the contractor's distribution of profit could he reasonably sustain a contention for extra payment in such a case. In any of the cases in this paragraph it is, as always, within the powers of the parties to agree to waive their strict position and thus for the contractor in particular to accept an unorthodox nomination, subject to agreeing any special terms that may be appropriate, and this is often done.

Before the settlement stage the quantity surveyor will be concerned with matters of tendering and examination covering nominated work, which will follow the general pattern of these activities for the main contract. Some nominated firms will have been selected before the main contractor and their names should have been stated in the main bills. Even if they have not, the contractor should be notified before his own tender is accepted, so that he may raise any objections that the contract permits and which he may have (see *BCPG*, Chapter 16).

SETTLING ACCOUNTS

In settling nominated accounts, the same general principles will run through as in the case of the contractor's direct work. The financial basis may involve quantities or some other form and will lead to the appropriate action. In those cases where a design element enters into the firm's work, the architect is particularly likely to be interested in the account; he should see at least the main documents, as discussed in the preceding chapter. In all dealings both architect and quantity surveyor should remember that they are to deal through the contractor, with whom the firms have their contract when once nominations have been accepted. This applies to all instructions, approvals and so on by the architect, and to negotiations and checking by the surveyor. If there is any relaxation of this funnelling in the interests of speed, it must be so delineated and watched that no lack of information or consultation results. Copies to everyone of everything are desirable. While the contractor is often only marginally interested in the financial outcome of a sub-contract, because of profit and discount, he will be far more concerned where there is a three-cornered dispute over who should pay for an item. He will see the account when it is eventually rendered to him, but it is desirable that he should be involved in all contentious matters and that his consent should be sought for all other direct dealings, making clear their scope.

With supplier accounts of the type that involve numbers of small items, as for instance ironmongery, the contractor and the surveyor will deal with the great majority of matters between them, and the supplier is likely to be involved only where unit prices are concerned; the contractor will have agreed the quantity for which he has to pay the firm, this being the total properly delivered, while the surveyor will agree with the contractor the quantity properly fixed. Any discrepancy—due, for instance, to breakages and theft—will be the contractor's responsibility. Special expense, such as unusual carriage, may need to be added in a supply account, while in the case of a sub-contractor it will be a matter of having a private arrangement between the firms in question.

Cash discount may lead to discussion in settlement if it was not clarified in the first instance. Should the contractor be asked to accept a nomination that does not afford him the correct level of discount in the first place, he must object at that stage; if he fails to do so he has no redress and loses his entitlement to the balance. If however a quotation offers some greater discount he may not retain the excess. This may seem an unfair position, but it is the state of things under the contract wording. A correction of the discount should have been made by a revised quotation or by an agreement to pay the contractor the difference in settling. Either way the nominated firm receives the same net payment, unless the contractor forfeits the discount quoted. Where the contractor himself incurs a special expense, such as carriage, this may not initially reflect a discount margin; it is equitable that an allowance should be added in the final account.

Fluctuations are straightforward to calculate under the JCT forms NSC/4 and NSC/4a, which are obligatory for nominated sub-contractors. Each form provides for the addition of one thirty-ninth to the net amount of fluctuations, to allow for cash discount. Under one of the alternative clauses in each form this produces an anomaly of principle by allowing in effect for two layers of discount when formula fluctuations apply (see *BCPG*, Chapter 18), not that this complicates the actual calculation unduly. When the addition of profit is dealt with in the main account, this is to be added to the whole amount of the sub-contract account including the fluctuations, by virtue of the provisions of the main contract payment clause.

In services sub-contracts in particular, there may arise the problem of sub-sub-contracts and these may be quite substantial. The position should be foreseen so that the nominated firm may quote inclusive of its own allowance for profit and discount. Where such a case occurs during progress, it is best dealt with by a negotiated addition of similar level, whether the result be expressed as a sub-nomination or as a sub-letting. Unfortunately no standard form of contract has taken this problem seriously in hand.

PAYING ACCOUNTS

The matter of direct payments to nominated firms has been referred to when discussing interim payments in Chapter 16. Where such a payment has been made to a sub-contractor because of the contractor's tardiness in paying, the payment will have passed the discount to the sub-contractor, since the contractor has forfeited it and the sub-contractor has become entitled to it; there is no reason why the client should benefit by it. The contractor is entitled to profit only on the sum that he actually pays and the final account should reflect this and thus give the client the benefit of this item, since he paid the sub-contractor. The contractor will still be entitled to his full payment for attendance, even if he has priced this as a percentage, since he has performed this function.

While the client has no general right to pay nominated firms direct, this may sometimes come about by amicable agreement, perhaps because the client has some special commercial relationship with the firm concerned. Here it is best to agree the terms for the adjustment when the special arrangement is made. If it is not, then the reasonable thing is that the contractor should be paid the discount and profit since he had waived his entitlement to make the payment, glad though he may be to be saved the trouble. He would however be going beyond the reasonable to ask for some greater margin when he had not raised the point at the initial round. If the client pays a firm direct without any agreement from the contractor, then the contractor may call the tune. Indeed, he may ask to be paid the full account leaving the client to sort out his double payment as best he may. Surveyor and architect alike will watch warily for such storms, to ward them off early.

In the main final account, it is usually sufficient to show the totals of nominated accounts, with an identification of the firm and the subject matter only. The detailed figures will be set out as a distinct variation account or remeasurement or whatever is appropriate to the nature of the account. It will be a matter of presentation whether it is necessary to split the totals between sections of the main account, as discussed in Chapter 17.

FORM GC/WORKS/1 AND THE ICE FORM

In dealing with a project related to Form GC/Works/1 or the ICE form a number of differences of detail arise over the matter of nominations and payments, which should be carefully considered if they are applicable (see *BCPG*, Chapters 27 and 28).

Fluctuations

The JCT form allows several distinct approaches to the question of fluctuation reimbursement or recovery: these embrace full or limited fluctuations dealt with by what has been called the traditional method throughout this book and what is widely called formula fluctuations, as it has been here also. Something is said on the virtues of using one or the other in Chapter 10, while the various clauses of the form should be consulted for the precise details of the items that are covered. Form GC/Works/1 has clauses of very similar significance in most respects. The ICE form also has corresponding clauses but they take more of a broad brush approach, giving a cruder settlement especially in the case of formula fluctuations. The general principles to be considered at the present level of generality are common whichever contract is in use (see *BCPG*, Chapters 10, 27 and 28).

SOME UNDERLYING CONSIDERATIONS

There are two broad courses open in calculating fluctuations, although a blend of the two may be employed. One is to make an adjustment of the contractor's expenditure on the project and thus relate fluctuations to a basis that is really prime cost; this is the traditional method. The other is to relate adjustment to the basis on which the contractor is paid for the work as such; this is the formula method.

As a basic philosophy the traditional method, as the first of these courses, is open to the same questions as are asked about a prime cost contract. These centre around the contractor's efficiency, which is not open to the same control by the architect in, say, a measured contract as it is in a prime cost contract. It may be urged that among the many ways in which a contractor may cause an excessive reimbursement of fluctuations are the following:

(a) Employing too many men.
(b) Working overtime and thus incurring non-productive time payment.
(c) Using men where plant would have saved their time and would not have incurred fluctuations.
(d) Performing work incorrectly and so incurring remedial time.
(e) Over-running a contract period, with or without an extension of time.
(f) Performing work late and thus running into a period of higher costs within the contract period.

Other such cases can arise. So far as items (a) to (c) are concerned it is fair to say that the contractor stands to lose far more than the client by any causes leading to extra payments, even if these are provable. Unless, therefore, some gross inefficiency arises that is beyond any bound of ordinary human error, it is reasonable to assume that the contractor has provided in his tender for a pattern similar to that existing on site and that this is his optimum balance. Even a degree of remedial work will be allowed for and may thus attract fluctuations—this appears as item (d) above.

Items (e) and (f) give rise to the same considerations as have already been discussed in dealing with prime cost in the preceding chapter; they may not however be treated in the same way under a JCT contract. In the case of late completion the level of reimbursement is specifically frozen at that at the due completion date, so that any further increase is not recoverable. A later amendment of the completion date will render it necessary to go back over any calculations and revise them accordingly. Since there is no power to affect the contractor's programme, it becomes almost impossible to do anything about late work within the contract period.

It will be clear that these distinctions are not entirely rigid and that the surveyor must use discretion. It may be a valid test for him to ask what would be the arguments to muster if a decrease were under consideration and not an increase. The contractor may find these same arguments relevant also.

The second broad course open in calculating fluctuations—relating them to the basis on which the contractor is paid for the work as such—may rely on quantities as part of the contract or may not. Either way, some analysis that employs measurement will be needed and is discussed under the latter two headings below. The relevance of points such as those listed as (a) to (f) above will vary. In this respect the philosophy of the two courses comes together to quite a degree; they are certainly affected equally by the considerations listed in items (e) and (f).

THE TRADITIONAL METHOD OF CALCULATION

It has long been a standard practice to deal with fluctuations by a direct examination of the contractor's costs. This requires the detailed checking of figures on labour records and invoices.

In the case of labour it is usual to ask for the contractor to supply at regular intervals—perhaps monthly—a statement for each week of working to which fluctuations apply. This will itemise the workmen by name and the various details of hours worked and other key data required. This is conveniently done by the contractor putting forward a carbon copy of the wage sheets with a suitable analysis of the fluctuations elements endorsed on them. Whatever the form of the statement, it should be signed by the clerk of works on the architect's behalf, as is done in the case of daywork or prime cost. Even if it is not possible for a visiting clerk of works to verify every single man on a full set of wage sheets it must be remembered that to falsify a complete set of sheets just to recover a fluctuations margin is hardly worth the dubious effort involved, unless the margin is large. Where only a summarised statement is put forward the position is open to rather more speculation.

When once the certified sheets are available the quantity surveyor will usually not need to adjust the hours for any reason, other than to deduct the hours worked on daywork if this has not been done already. These are usually allowed at current rates for daywork and thus no fluctuations arise.

For materials the usual basis of adjustment is by consideration of the invoices on which the contractor has paid for the items. It is uncommon to require any signature of a record of total deliveries; usually the invoices will record the point of delivery and thus establish that the materials are relevant to the project. Two exceptions are the cases of materials delivered to the works of the contractor or a sub-contractor, and of materials supplied by the contractor from his stock in his yard. In the first case the date of delivery will often give a reliable guide; in the second a signature from the clerk of works should be obtained. With materials it is easier than with labour to make an overall check on the quantity given, by an analysis of the measured account with a reasonable allowance for waste. Again the total used in daywork should be deducted.

While the quantities of materials are thus easier to determine, the determination of what prices to adjust is more trouble. Where 'full' fluctuations apply the market price is the standard; the establishment of the basic price list has been dealt with in discussing the examination of tenders in Chapter 11; it must be checked that changes in price are really due to market fluctuations and not simply to a change of supplier. Where there is a general move of the price of a commodity the position is not hard to establish, so long as an individual supplier's charge is in step with it. Where a single supplier (as is perforce under consideration in a single case) makes a change all on his own there may be a case for resisting an increase in particular. The standard requirement that the contractor must give notice of a fluctuation is most valuable in dealing with materials, so that the situation just mentioned can be forestalled by a query or a change of source. If materials are supplied by the contractor from his own stock in small amounts to cover localised shortages, this notice will usually be waived and a current price may be used if a change is applicable. Where a large amount is proposed from stock, special agreements should be reached beforehand. In all cases it is necessary to ensure that adjustments are related to prices for full deliveries, to materials at the correct stage of processing or conversion, and to the proper level of discount. All these matters should have been made clear in the basic list and its attachments.

If, however, 'limited' fluctuations apply a different approach is to be used. While the market price shows that a change has occurred, it does not say why; hence it is necessary to analyse any change to see what part of it, if any, is due to statutory action. A supplier may state that an increase is the result of fiscal changes, but this may not be an exhaustive explanation of what has occurred. In view of the scope of taxes and so forth that may be relevant along the whole chain of supply and the delay in their effects coming forward, this whole matter calls for a fairly broadminded solution by surveyor and contractor alike.

THE FORMULA METHOD OF CALCULATION

Discussion in Chapter 10 on the decision to use formula fluctuations and the basic elements of the approach, together with discussion in Chapter 16 on timing and accuracy of valuations, give a sufficiently broad introduction to the subject for this book. The reason for precision in these matters of timing and accuracy is that the final calculation of formula fluctuations takes place at the time of interim valuations, except for 'topping-up' amounts at final settlement. Under the contract and the formula rules it is not possible to go back and adjust calculations once made, except for amending errors in using the formula apparatus. Any error in the quantities of work included must stand and be dealt with under the topping-up arrangements, which means that the final figures calculated will almost inevitably be somewhat different from those arising if there is no error.

The quantity surveyor has to prepare sectionalised or annotated bills and sectionalised valuations, enter the resulting totals on the pro-formas that are available (or spirit up his own) and apply the published index numbers in accordance with the rules. In principle he is adjusting the measured rates by substituting a series of values for the unknowns of a formula to arrive at a total. In detail he is following a number of fairly intricate rules in the case of the building formula to cover many of the finer adjustments that arise out of reality, while in the case of the civil engineering formula he is following fewer rules of a cruder nature. But even in the case of the building rules it has been argued in Chapter 10 that the method is a fairly blunt instrument in that it uses national index numbers and thus ignores regional variations and the working methods, efficient or inefficient, of the individual contractor. Against this the weaknesses inherent in a method related to the contractor's actual costs have already been considered.

While calculation is in principle automatic if there are no snags during progress, there are special provisions to cover delays in the issue of the bulletins of index numbers and again delay by the contractor in completing the works, as under the traditional method. In the former case these lead to special calculations, either on a temporary or perhaps a final basis, to bridge the period until the bulletins reappear.

A MODIFIED METHOD OF CALCULATION

The standard forms are quite open as to how the traditional method is to be used in calculating the amounts of fluctuations and do not set out any requirements such as the intensive use of time sheets and invoices. Indeed, the JCT form, in allowing the surveyor and the contractor to agree a final figure due to any particular fluctuation, appears to be hinting that some less laborious process might be employed. This is reasonable enough, because the labour of calculation in the older approach is often out of all proportion to the result. Any alternative must rest on what the contractor is being paid rather than what he is expending.

While this means adopting something akin to formula fluctuations, it is impossible to import that method wholesale into a contract that is not already based upon it. To do so would be to change, among other things, the scope of items subject to adjustment and the base level from which they would be adjusted. What is conceivable is to extract the overall labour and material content of suitable work categories from the final account and to divide this into the time periods of the contract in which any differing prices applied. This division will be related to the totals in the valuations, adjusted for any tendency to be over or under at any stage that is revealed by the final figures and perhaps also viewed in the light of the clerk of works' records and the construction programme as achieved. The actual price changes per unit, once the total is calculated, will be arrived at on the same principles as are considered for the traditional method, as far as these are relevant, and applied to the totals on a percentage basis to give the fluctuations adjustment.

This type of approach, when worked out in detail, means taking the prices in the individual contract to pieces to arrive at the equivalent of some of the data underlying the national base index numbers of formula fluctuations and then using the actual changes that have occurred on the contract, which in turn are the equivalent of some of the data underlying later index numbers. The results are applied to the still fragmented prices, rather than to the complete prices as with the formula method proper. Using this approach for a whole contract of any size and duration may be considered too lacking in accuracy (though perhaps no more so than civil engineering formula fluctuations), and if the calculations are over-refined for the single contract they may take longer than the traditional method. However there is no reason why some such modified method should not be applied to a section of fluctuations only, while the traditional method is used elsewhere. A direct attack on the wages sheets will still be far easier for, say, a tool money increase affecting the last few weeks of a contract only. So far as is practicable a consistency of method is desirable, but a balance should be kept with realism.

Daywork

Little need be said here about daywork. A great deal will be determined by the rules embodies in the contract; thereafter the quantity surveyor is dependent on signed records, as discussed in Chapter 17. The valuing of daywork affects traditional fluctuations as has been stated above. A daywork sheet is a miniature prime cost account and some of the points outlined in Chapter 18 may occasionally have a bearing when dealing with daywork. Since daywork is related to work carried out on another basis, it is always necessary to ensure that the two types of payment do not overlap; the reverse is not likely in the nature of the case.

Value added tax

While construction work is not currently exempt from value added tax, most work other than alterations, maintenance and certain equipment is zero rated at the date of writing. It is thus necessary to single out those items that attract VAT from what will usually be the bulk of the account. The standard forms vary in their arrangements for dealing with the tax, the JCT forms providing for the whole matter to be dealt with under a supplemental agreement and for payments to be made separately from the contract amounts, while both the GC/Works/1 form and the ICE form include matters within the contract.

Either way the detailed work will be much the same, although there are some fine distinctions. While the contractor is responsible in principle for providing detail as the collector of the tax, the surveyor will need to check the sums or perhaps in practice perform some of the calculations. The forms include various provisions about disputed amounts and the like, most of which take their substance from the requirements of the statutory regulations (see *BCPG*, Chapters 13, 27 and 28). These regulations must be carefully followed if affairs do run off course.

CHAPTER 20

Payment for unexpected loss and expense

Within times that are comparatively recent — as things move in the construction industry — 'claims' have become quite respectable. The reasons for this are complex and of little immediate concern but some of them may be mentioned in passing. Contracts are becoming larger and are themselves more complex, while time-scales are shrinking. Some ways of going about work from the client's end mean that there is less real system, although the client may be prepared to pay for this, as discussed in Chapter 5. Architects are losing their aura of autocracy and quantity surveyors sometimes go against tradition by being in the direct employ of the client. The contractors have become less tolerant of insolvency or more desirous of a sure profit, and even have surveyors of their own to plead their case. Some people may even want to place the credit or blame for current trends on the permissive society.

In some circles it is said that if claims are of such frequency there is little point in having bills of quantities or, for that matter, competition. All that is being done, it is alleged, is to ignore the bills, calculate the prime cost in some rough form, and then pay the contractor on this basis, but making it look as though he has been paid some amount that is 'extra over the bills'. Even if this has happened in some cases, it is not necessarily a sufficient argument for not seeking some form of fixed basis in the first place; the contractor always has an incentive to economy in that he may not receive his cost in the end and thus should keep it within bounds. Where a claim is genuinely dealt with on an 'extra over' basis as it should be, then the bills have been even more effective as a sheet anchor on the way along. More will be said about the calculation aspect later; first the wide term 'claim' must be pared down.

The scope of the subject

Many of the discussions that surround the agreement of a final account are outside the present scope. They will be concerned with points of a technical nature relating to the measurement of items and, even more often, to their pricing. These discussions may proceed on a high plane and may involve large sums, and if they reach deadlock they may result in expensive legal action or arbitration; but they are not claims of the type covered here. They will usually be about variation items, although they may also embrace omissions from the documents, particularly the bills, or ambiguities within them. With some ingenuity they fall to be dealt with within the formulae set up in the contract.

An unexpected loss or expense as discussed here means one caused by those matters set out in JCT clause 26.2 (see *BCPG*, Chapter 7) and also the results of any wider breach. They are thus expenses that would not have been provided for in the tender, because they could not have been known about at the time, even if the documents had then shown the physical items, if any, over which they later have occurred. They have in common that the work looks just the same when it is finished as it would have done if the loss or expense had not occurred. There is quite a limited list of causes, which must have had the effect of disturbing regular progress of the work, or some part of it, to a material degree. These causes are all matters for which the client is responsible in the sense that they are either his or his architect's act or omission. They include changed access, incorrect information, failure to provide information on time, postponement of work and some cases of introduction of direct contractors by the client; these items may therefore have a widespread effect on the works and may also lead to extension of time.

Among the causes are architect's instructions over expending a provisional sum or over a variation (see *BCPG*, Chapter 9), which latter is defined in clause 13.1 as a physical change in the works or as a change in particular restrictions in working conditions defined in the contract documentation. These items are thus different from most of the other causes given, in that they lead in their own right and in the normal and even course of events to an adjustment in payment to the contractor. All of this straightforward adjustment to allow for the localised effect and cost of any item is governed by later parts of clause 13 and is discussed in Chapter 17. When these items are instructed in such a way or at such a time that there is disturbance not contained within the work so defined or varied, then it is the intention that the loss or expense should be covered under the claim procedure of clause 26. There is the particular possibility that a number of variations, each of relatively limited scope and impact, may together produce a disturbance of sufficient consequence to lead to a claim.

In the case of any variation there may be difficulties in separating out precisely the cost of the variation and the amount of loss or expense, clear though the distinction may be in theory. When the variation is a change in a restriction of working conditions, even the theoretical distinction may not be clear and the sensible policy then is to deal with the whole cost as being due to the variation. It may even be that the net effect is a reduction in cost on occasions.

The contractor is not bound to accept the framework of the clause in reaching a settlement and may still go to arbitration or to court either in addition to, or instead of, applying under the clause. In practice the parties may well choose to settle claims arising out of other breaches by the client in the manner that this clause permits, although neither party is obliged to do so. Further if the contractor agrees to bring forward the date for completion, or to do something else which he is not forced to do, he may be paid for the expenses involved in this on a similar basis, unless a consideration is agreed in advance. That he is to be so paid should, however, have been agreed before any payment can be made, and for his own certainty it should be agreed before he takes any action.

Form GC/Works/1 covers the ground of the JCT clause in its Conditions 9(2) and 53 (see *BCPG*, Chapter 27). Instructions have a much wider reference, however, than the list in the JCT form. The form also contemplates a claim-in-reverse by mentioning the recovery of any saving. In Condition 44(5) a special case of payment on a hardship basis for a determination not due to the contractor's default is given, although this is distinguished by a lack of any right of appeal to arbitration or the courts. This clause is virtually a claims clause for the present purposes, since it gives no formula for settling the amount.

The ICE form is rather less precise than either of the others in detailing the causes of claims. It places emphasis on advance warning of intention to claim and the rendering of statements of claims, but does not make these actions conditions precedent to the right to claim (see *BCPG*, Chapter 28).

What none of the standard forms does is to allow the contractor to claim on grounds of losing money for reasons undefined or unknown, where the root cause could be low tendering or bad organisation. There are also risks of commerce or physical conditions which the contractor may be carrying and these should remain with him in terms of their effects. While one contract may suffer by them, another may benefit equally from them. There may occasionally be some quite exceptional circumstance in which a client feels that he has a moral obligation towards his contractor, but the surveyor should not hesitate to advise either party as to their contractual responsibilities at any time. If a client does decide to admit a particular claim from a contractor on a purely *ex gratia* basis, the surveyor should evaluate it on the same principles as any other contractual claim, while observing carefully the boundaries within which the client has offered to make payment.

The contract procedures

JCT clause 26.1 requires the contractor to make the first move by an 'application' properly given in writing 'that he has incurred or is likely to incur' loss or expense. This he must do as soon as he is aware of the disturbance of progress, actual or potential. This allows proper records to be kept of what may be a more complex and extensive disturbance. But it may also give an opportunity for the architect or the client to take avoiding action to mitigate the effects and thus cut down the cost. This is a delicate matter, in that the contractor's organisation remains his concern, disturbance or not, but where the cause is a continuing one it may be possible to deal with it.

Without a notice from the contractor nothing can be done formally by the architect or the quantity surveyor, although there may be cases in which it will be reasonable at least to hint to an uninformed contractor that he may be entitled to claim. It is certainly prudent for both of them to keep notes of events if the contractor has not set matters in motion, but a claims situation looks like arising. The period within which affairs should become apparent to the contractor is not always easy to fix with precision and, even if the contractor is clearly late and so has lost his right to operate the procedure of the clause, he may still be able to use his 'other rights and remedies' as the tail of the clause recognises. Too much information is always better than too little.

The architect is named as responsible for determining the amount of the loss or expense, unless he chooses to refer it to the quantity surveyor, as he usually will. The naming of the architect is unfortunate in one way, since he may have been the person whose inaction or error led to the claim in the first place. He is thus cast in the joint roles of judge and defendant. He must produce a section of the evidence; beyond this it is as well if he leaves the financial outcome to the surveyor. The employer is not named in the clauses, but there may be cases in which he will be directly concerned in the fact-finding process. Indeed it may be best to obtain his agreement to the amount arrived at for any major claim at the time, rather than to leave it to surface in the final account. Claims are naturally rather emotive and a gradual appraising can only help. Either party has of course the usual right to go to arbitration over his part of a final account if aggrieved.

All this having been said, it is in practice common for the contractor to assume the onus of proof over the claim. He is required to demonstrate the fact of disturbance and supply information upon request for the architect's adjudication under the contract. He will usually produce a financial assessment as well, rather than leave it to the architect or the surveyor to do so. This is a rational approach, since the contractor has the effects of the disturbance immediately at his finger tips, while the others would have to start the exploration from without.

Clause 26.5 and clause 3 read together allow 'any amount from time to time ascertained' to be included in the final account and in interim payments, where there will be no deduction of retention. This clearly means amounts formally agreed, but further clause 3 allows additions 'ascertained in whole or in part' to be included in interim payments and this, read with 'from time to time', reasonably includes approximations in advance of final agreement (see *BCPG*, Chapter 4).

Form GC/Works/1 broadly follows this pattern. In the case of instructions it requires these to be in writing before any claim can be raised based upon them. Thus the contractor must be on his guard to seek confirmation. If the contractor fails in this procedure it is common practice for his eventual claim to be treated on what is termed an 'ex-contractual' basis, or even an 'extra-contractual' basis. These quite un-legal expressions mean that the superintending officer does not have direct power under the contract to deal with the items; they are usually referred up the line of departmental organisation to be dealt with by others. Since the causes arise within the contract, they will not be ignored as supporting a plea for reimbursement, despite the term 'condition precedent' used in the contract. When a claim has been agreed, but not before, its value may be included in advances on account but will be subject to the reserve.

Nothing need be added here to the brief reference to the ICE Form under the preceding heading. In what follows, it is not necessary to distinguish between the forms in dealing with broad principles. It may be convenient, in any negotiations that range fairly broadly over the history of the contract works as performed, to take into account any related matters, such as liquidated damages or other counterclaims. While they should be formally cleared under the appropriate procedure of the contract, they inevitably will influence the level of settlement reached and cannot therefore be considered in isolation.

While the early statement of a figure for a claim is not essential, the temptation to put off its computation should be resisted. It may depend in some cases on what is included in the rest of the final account, but otherwise it should be put in hand as soon as possible so that it does not itself become a source of uncertainty. In particular, matters that properly belong in the normal variation account should go there and not be deferred as part of a claim just because the two surveyors are avoiding the issue of agreeing rates. There is a strain of contractors, fortunately rare, that appears to consider that the wafting about of a claim in principle, without any supporting calculations, creates some psychological atmosphere that will gain them favourable consideration in the meanwhile on other issues. Here the right of the architect, under the JCT forms at least, to ascertain the amount should be pressed to end such a situation.

Sources of data to be used
THE CONTRACT
What will need to be collected together to evaluate a claim will depend on the nature of the claim. The starting point must be the contract documents.

Within the contract conditions there may be named the specific cause that has led to the present expense; if so, the rules for settlement given may define some at least of the information that is required, as well as the scope of reimbursement that may be permissible. Of the remaining contract documents, the most likely to be of assistance are the drawings and the preliminaries, since the technical sections are more likely to be related to points that are amenable to a measured or similar valuation. Such causes of trouble as a change in the scope of the works or their character, or the freedom of working conditions available on the site, will often show from these documents. So too may an intended sequence or some intermediate date acting as a programme restraint. Any correspondence at the time of examining the tender may also shed some light, although it must be read as outside the formal contract and thus of value in interpreting the intentions of the parties where the contract is silent. If there is a likelihood of arbitration or legal action following on a breakdown of negotiations, then this supplementary evidence may need to be viewed more critically for formal value.

This initial survey will be made to determine the intention of the contract, which is now open for adjustment. It may lead to the conclusion that the contract has been effectively subverted in some way by the actions of the parties, so that a reassessment of the whole basis of payment is called for—possibly but not necessarily on a *quantum meruit* basis; this is a drastic state of affairs and quite beyond the scope of the present discussion. It is also beyond the powers of the client's advisers to take it upon themselves to agree to such a reassessment without the client's agreement or even to give authoritative advice that it is legally inevitable. They can go no further than to suggest that things have reached a desperate pass.

THE CLIENT'S ADVISERS

Assuming that a less radical claim is in the balance and that the quantity surveyor has been asked to deal with the whole matter, he will be responsible for gathering evidence of what has happened from the various persons who have played any part in the drama. He may well have some fairly strong views of his own on what has happened, from his position on the touch line as he deals with the accounting function. Indeed, if things have been building up to a claim he will have been making his own records of the effects from those beginnings, even though the contractor may not have given any formal intimation of claim. These records may consist of little more than his personal impressions in some cases of the relative efficiency of the others involved, but they may be very useful if the stage is reached at which the surveyor becomes a father-confessor sifting the evidence and affirmations that the others produce. The ability to remind them tactfully of the real interaction of certain cold facts may be just the catalyst that leads to a conclusive reaction, preferably without too much heat. There is no substitute for adequate documentation, but this may not be enough in itself. Disturbance is in part an individual phenomenon and some persons and some organisations are more susceptible to it than others in similar circumstances. Whether they always merit reimbursement on account of their high vulnerability is another matter to be decided in the particular case.

Which of the other persons is to be approached first will depend on how things have developed. If the claim has been originated in the way that the standard contracts envisage then it will reach the quantity surveyor through the architect and he should be asked to supply any further information that he has not already put forward for guidance. This will include his direct comments on the effects that the contingency has produced so far and, if still running, is likely to produce. If he has issued an extension of the contract period this information should have been supplied already; if it has not been it should be obtained or at least an initial forecast of its size extracted for preliminary guidance. In terms of pure fact, the architect should give a schedule of dates relating to the claim history. This may include his drawing register, supported by the comparable dates on which the contractor made written requests for the drawings. Variation and other instructions should be treated in the same way, with a particular eye to any postponements. Alongside these, a statement of the physical progress of the project at the time of issuing any of this information will be needed. If nominated firms are involved, the dates of approval of their designs may be relevant. Any general correspondence with the contractor and with nominated firms (through the contractor, it is to be hoped) that reflection suggests to be useful may also be requested.

Others on the client's side of the fence who may need to provide information are the clerk of works and any consultants. In the first place, the architect should collect their statements to append to his own. It will be a matter of protocol, tempered by expediency, as to how far all subsequent discussions are channelled through the architect; he has the overall picture in his mind, but these other persons may see some aspect in a fresh and important light. Again, direct communication is always quicker than indirect; this point is probably the most cogent one to make to a hesitant architect, especially if it is reinforced by an outline of the time consumed by constant use of formal channels. When these persons have been consulted obviously the collated results must be discussed with the architect in the light of his overall knowledge.

A particular source of data that these persons may have available lies in their diaries, especially for site visits and reports. They may also be on record in the minutes of site meetings, and what was there recorded in miniature may need to be expanded but not embroidered for present purposes. Minutes present two problems of their own so far as their use goes. One is that they do not possess the status of written instructions or even of written requests from the contractor; here they must be taken if needs be as the best available, or perhaps as a way of reminding the architect that a confirming instruction is due. The other problem comes first in time and is evidential; usually minutes will originate from either the architect or the contractor and it is a commonplace that two persons' statements of the same facts can vary quite amazingly when seen by an interpreter. The origin of the minutes cannot be ignored, especially if they were produced in a claims-conscious atmosphere.

If there has been a direct interference on the part of the client, it may be necessary to interrogate him also. It will again vary from case to case as to whether to make this approach entirely through the architect. The quantity surveyor may have been appointed direct by the client and thus have complete freedom of access but even here, in the interests of co-ordination and in recognition of the peculiar position of the architect in the team, it may be far better to operate through him to the same extent as when dealing with a consultant.

In all these cases, the surveyor must seek to maintain his detached position and where there has been a deterioration of relations during the contract, as can occur in these circumstances, he must act with great care in all dealings. This is in the interests of all concerned; it may become very much in his own interests if he desires to have any continuing relationship with those who may be the source of his future commissions or promotion. But if it comes to an issue, he must be ready to hold to what he sees as the right answer in the face of the attitudes of those who may have their own vested interests at stake. He may even remind his client that he, like the contractor, has arbitration available if he does not accept the amount assessed in the final account!

THE CONTRACTOR

It may be that the first person with whom the surveyor has any dealings over a claim is the contractor, if only because he has been regaled with a series of tales of woe at every site visit. These will have alerted him to the situation, although he will have done what he can to probe their cause. The contractor may be having difficulties within his own organisation or he may just be one of these who pitches everything on the black side to produce a sympathetic attitude in the quantity surveyor, a tactic which may have quite the reverse effect if overdone. But there is not often smoke without fire and the philosophy of the contract forms is that the contractor is going to receive a straight run at the job for the contract price. This being the case, the surveyor must receive the contractor's representations seriously and in the attitude of seeing both sides of the problem. What, however, he must be cautious about is taking any unduly sympathetic stand during any informal chats of the sort that inevitably develop on the site. Otherwise he may find himself in a false position later.

The major contribution that may be expected from the contractor if he is on his toes will be a formal statement of the amount of his claim. This is sometimes embodied in an imposingly produced document. As has been said, this is not a requirement of the contracts, which make the architect or his equivalent responsible for arriving at a figure but it is a likely way for things to develop, unless contractor and surveyor sit down together to produce a result. This, like the similar approach to negotiation, may be the most straightforward and avoids the common situation of an inflated claim followed by an inevitable cutting down process. It is perhaps harder to achieve in the inevitable circumstances.

What the contractor must be prepared for is the disclosure of what would otherwise remain his confidential information contained in the relevant parts of his cost records and so forth. This may mean the basis of his bonus arrangements and his overheads, where the claim warrants such a searching approach. An auditor's certificate may be desirable to establish overheads.

Principles of the calculations

It may be discerned that there is a jump in the sequence of the narrative at this point. What follows relates to how the figures in the claim are handled once they are to hand and once all the detailed arguments have been adduced for and against the contentions of the various persons. This is deliberate and even inevitable. Claims differ widely in their causes and the interaction of causes, and in the effects that they produce. The JCT forms blandly require the architect or the quantity surveyor to 'ascertain' (a very precise term) the loss and therefore to assess the causes with commensurate certainty. This is not an area where such precision is attainable and there is, alas, no acceptable formula into which the various figures may be dropped to come out at the end as a predetermined amount. In the nature of things they are the product of what should not happen and may give rise to a multitude of problems in their settlement. A number of general cautions may be set down here.

THE 'EXTRA OVER' PRINCIPLE

A claim is a contractual request for reimbursement of the additional expense not envisaged, and therefore not provided for, in the contract sum and any rates supporting it. As such, it should be expressed as an extra sum or sums in relation to the original basis. Where a claim is in respect of a substantial amount reflecting serious disruption the contractor may be tempted to calculate it by taking the amount that he is otherwise to be paid under the final account for the section of work affected and deduct it from the cost to him of performing that section. This over-simplifies the issues by making the assumption that the original pricing was at precisely the level that the cost would have been without the disturbance and that the whole of the difference is due to it and not to any change in the contractor's basic efficiency.

Where it is the whole cost that is set against the whole account, this procedure is suspect enough and only justified when a *quantum meruit* payment is being conceded. This is not being considered here. The procedure is even more suspect when only a part of the finances is subjected to this treatment; here there is the additional possibility of an overlap with other sections of the work not affected by the disturbance. This is a familiar state of affairs when dealing with daywork. For these reasons the evaluaton should proceed from the premise that the contract stands and that only the effect of the disturbance is being calculated.

When this has been said, it remains true that in any case where a great part of the works has been affected it is usually a very real assistance to have a knowledge of how the overall costs incurred by the contractor have been spread. This knowledge can act as a regulator in assessing the reasonableness of figures put forward or of the surveyor's own calculation. For example, a particular item of claim may centre around the provision of an excessive amount of transport and be evaluated at, say, £4,000. If the total cost of transport is only £6,000, this would suggest that the contractor has got away very lightly with his basic cost of providing transport or that something really catastrophic has occurred in relation to the claim area. An analysis of the original contract amounts to discover the sum allowed for transport may not have been at all conclusive, while this cross-check will help considerably, even when due allowance has been made for the considerations raised in the last two paragraphs. A large amount of such information will already be in the surveyor's hands where fluctuations are being calculated by the traditional method and the rest should be obtained.

HEADS OF CLAIM

Whenever a claim is related to several causes, it should be presented if at all reasonable, under divisions that reflect these causes. If there has been uneconomical working at more than one stage this should permit a clear segregation, unless some other part of the work was so delayed as to affect one of the later stages. Even where the division is not absolute, a fairly clear split of the effects should be assessed. Only thus can a relationship with the contract basis be adequately maintained, not to allow formal consistency but, be it repeated, to assess the claim at the proper level at all.

How the figures are calculated and shown in the final account will depend on the nature of the claim item. Sometimes it is suitable to go back into the contract rates and analyse these so that percentage or other increases on their contents may be ascertained. In other cases it will be better to go out beyond the rates and assess the loss or expense as an independent issue, that is to say, upon the equivalent of a star rate basis. Whatever the approach, or combination of approaches, that is used the result may be shown in the account in whatever detail is suitable. If the process just outlined is as simple as it has been stated as being, then virtually all the detail may be put into the main account. If it is a complex as it often is in fact, it will be best to produce the figures separately and include the totals only in the account.

A somewhat suspect method of building up a claim is what may be termed the 'compound percentage' method. It may be contended that cause A has led to a loss of production that amounts to x per cent for some part of the works; however, the result of cause B has been that all the hours worked on the site were subject to a loss of y per cent, including those worked to make up the deficiency due to cause A. It can be seen that this is a very difficult phenomenon to observe on site and that the person advancing such an hypothesis is under a prima facie obligation to substantiate it. There will be circumstances, such as prolongation, that may lead to something like this situation, but the mathematical reasoning of the calculations should always be examined very closely.

PARTICULAR ITEMS
Some costs can be a more frequent seat of trouble than others. Usually materials are fairly straightforward, since matters seldom extend beyond uneconomical purchases and other factual considerations. Formwork may enter into a claim rather than be part of a pro-rata price adjustment, but should be dealt with on a similar basis.

Preliminaries may need adjusting to cover prolongation and this will lead to an analysis of the contract items on the lines dealt with for interim payments, but with more attention being paid to the detail since the figures have a final significance. If the preliminaries have not been priced fully in the bills, the possibility of a special assessment may arise. Where there has been an acceleration of progress, or some other intensification of effort involving additional supervision, a corresponding calculation is needed, although the figures may be employed rather differently in detail. It will not always be possible to make such a proportionate adjustment of preliminary items, particularly if supervision is affected. Sometimes it may be the case that an extra foreman was drafted in or that the area contracts manager spent much longer on site than he otherwise would have done. In the former of these two examples there is a clear additional cost; in the latter it may fairly be asked whether an extra cost has arisen or whether some loss has been suffered due to the manager giving less attention to other sites. The latter awkward item is made doubly so by the term 'direct loss' as used in the JCT form, this being subject to quite rigorous legal interpretation.

Where measured items are subjected to any analysis, the labour content may be scrutinised to look at a drop in production and also the associated bonus that was not earned. It may be contended that it still had to be paid, to retain men who were frustrated by matters beyond their control. Here the realism of the targets set may need examination. If all the work carried out between certain dates is held to be subject to disruption, there will still be the daywork to be deducted from the whole; daywork does not suffer from low productivity as far as the contractor is concerned.

Plant has its peculiar aspects, since it is both productive and inanimate. It may thus be left standing on site without the reason for this being crystal clear. Here is a very definite case in which it is unsafe to take cost as the main criterion in assessing the loss. There just may not have been anywhere else to place the plant if it happened to be the contractor's and not hired. Even when a chargeable duration has been fixed, there will remain the question of an appropriate rate for the period. A wholly-owned plant hire firm, or an internal plant department, may both charge high rates for reasons of group or company policy which are no responsibility of the client. The national plant hire schedules are also pitched high, as they are intended for incidental plant items on a daywork basis. A rate deduced from first principles may be best, although where the item is known to be allowed for in the preliminaries the allowance there may serve as a basis for the further calculation.

In the case of nominated sub-contractors' claims, the general approach is similar to that for the contractor's claims. The contractor is required under main and sub-contracts to pursue claims on behalf of a sub-contractor, who has no direct access to the client. This is not intended to apply to claims arising out of the contractor's default, and it is necessary to look closely at any sub-contractor's claims to eliminate events of this nature from the contractor's main claim. It may be expected that the presence of claims with multiple causes will lead to involved discussions over segregation. An addition must be made to any sum so that the sub-contractor can allow cash discount without eroding reimbursement of his direct loss. As the gross sum will then be included in the sub-contractor's final account it will rank for addition of the profit percentage due to the contractor, without the doubts raised in the next paragraphs.

The best wine has been kept until last and is in two bottles, respectively labelled 'interest' and 'profit'. Where the settlement of a claim is substantially delayed, other than by the contractor's failure to come to terms or provide information, it would seem reasonable that he should be allowed interest on the sum outstanding. It has often been said that interest is not customary in the industry, but customs change and in any case interest would be allowable in a judgement debt; there is a precedent in the ICE form in that interest is chargeable on overdue interim payments. Several recent decisions in the courts have allowed financing charges and amounts for inflation during the delay period. These may not be universally applicable, but do indicate the trend (see *BCPG*, Chapter 7).

Profit is a more difficult matter; it is clear that the level of claim settlement should meet the actual loss suffered by the contractor within those fields recognised by the contract or by the wider law. This will mean that his original profit margin is no longer eroded by the loss and is reinstated. It has been strongly argued in legal circles for many years that although the contractor has incurred extra turnover that he did not ask for, and might well have looked for in the shape of fresh contracts, he is not entitled to receive a profit on the reimbursement of his loss. He is to receive what he has lost as of certainty and not what he might have gained on other but uncertain and unobtained business. This is a hard line, but alongside it may be put the argument that if the present contract were running at a loss, quite apart from the claim settlement and related causes, it would be clearly unfair to make a deduction from the settlement to allow for a proportionate loss on the settlement. Nevertheless, some of the cases just mentioned have also allowed profit and suggest a change in the judicial view of the commercial world. It may also be surmised that there are occasions in settling claims when the agreement of a profit margin (even if it be termed '*ex gratia*') has come about as part of a compromise to avoid the possibility of an arbitration or worse. And not many contractors are going to quibble over a name so long as the cash works out the same.

There may be other overhead expenses that arise directly out of the disturbance. If so they may be allowed, whether they are a site expense or one at head office, so long as they are genuinely extra. A contention that additional turnover must automatically mean additional overheads cannot be supported. There may be no additional overheads at all, although on the other hand particular turnover may lead to disproportionately high additional overheads.

CHAPTER 21

Damage to the works during construction

As a basic legal principle the contractor, under a normal building contract, agrees to deliver up the works in their proper condition on the due date (see *BCPG*, Chapter 4); so that if there is any damage to them that is not otherwise treated in the contract he will be liable to make it good at his own expense. The standard forms all erode this principle to some extent; Form GC/Works/1 does this by its policy over accepted risks, while the ICE form has provisions for excepted risks which have a similar effect in a narrower way (see *BCPG*, Chapters 27 and 28). The JCT forms also do this where clauses 22B or 22C are used, since the client not only insures but also takes all the risk (see *BCPG*, Chapter 8). In all of these cases the client pays the contractor for the cost of the reinstatement; how he then recoups his loss is not a matter with which the contract concerns itself. He may turn to the taxpayer or the ratepayer, or he may have an insurer; in this last case it is quite possible that the quantity surveyor will be drawn in to negotiate a claim with the insurer on behalf of his client, on whatever basis may seem best. It may be suitable to agree to a lump sum settlement; if not, an agreement related to the payments made to the contractor may be acceptable. In either case the procedure is simple in essence, although some of the issues raised below may have a bearing on it.

The other case under the JCT forms is that of clause 22A, under which the contractor takes out an insurance for a restricted list of contingencies in the joint names of the employer and himself, while retaining the risk. Alternatively, and more commonly, the contractor will arrange for the employer's interest to be endorsed on the running all-risks policy that he is maintaining for his contracts and other work generally. The architect has powers to ensure that this insurance has been effected and is being maintained from time to time, or to take alternative action. Under this clause the contractual and related procedures are more complex.

The rest of this chapter is concerned almost entirely with this position and with the actions that follow upon damage occurring.

The employer's position

Although the employer is named as insured, he will usually assume a passive rôle in the settlement. This is because the contractor stands to gain or lose most under the financial negotiations and may therefore be expected to pursue the claim with the maximum enthusiasm. All that the employer is due to receive out of the claim is any percentage that has been included for professional fees. To this extent he has an interest in how well the contractor makes out, but can hardly influence matters when he has such a marginal share to come. What he has to expend on fees in any case may be quite a different sum from what he receives, since much will depend on the nature of the work to which his advisers are put. The contract provides that the contractor will be paid for the reinstatement work on certificates of the architect; this will mean establishing a joint holding account for the monies as received, unless it be agreed to draw them from the insurer progressively.

Apart from this the employer can do little more than urge a quick acceptance of the claim, since the contractor is not required to commence the reinstatement until the claim is accepted. The employer is due to concede an extension of time and he will receive no compensation under the policy for this delay, although neither may the contractor. In passing, it may be mentioned that clients should be advised of the virtues of a policy that will cover this type of delay and the resultant loss. If a total suspension of work for a prolonged period comes about, there is even the possibility of the contractor determining under clause 28, if he is so inclined.

The only situation in which the employer is likely to be thrown right into the arena of settlement is in the event of the contractor becoming insolvent in the middle of reinstatement. Here the employer, as jointly insured, would be able to take up the claim directly so that the monies could be applied to the costs of reinstatement under the new contract.

The architect's position

The primary responsibility of the architect is to ensure that the employer receives the works for which he contracted, without extra payment, since the contractor is still responsible even though he has the insurance to fall back on. He is also concerned that there is no danger as a result of the incident for which his client has any responsibility. His most pressing action must therefore be to check on the stability of the works—although this, again, is the basic concern of the contractor; otherwise the indemnity provisions of the contract could well lead to the employer meeting a claim in some circumstances. Any work that was of an emergency nature could safely be put in hand without formal sanction from the insurance company, who would regard it as mitigating the damage.

Subject to any first aid action of this type, the architect may embark on his assessment of the damage in full. Here he will also be in close liaison, in any serious case, with the building inspector or district surveyor, as the case may be. Strictly, he should instruct the contractor on the work to be done and leave it to the contractor to seek the appropriate level of payment from the insurer but in a case of any complexity it will ease work for everyone, in the long run, if the architect meets the assessor and confirms the scope of work directly with him. Should the assessor not be prepared to admit any item in discussion with either the architect or the contractor this does not prevent the architect from requiring the contractor to perform it. The contractor's remedy over this, as with any other disagreement with the architect, is to seek arbitration against the employer.

Once work recommences on the damaged section the architect exercises his normal functions of supervision, during which he will still retain the right to order any additional work that may become apparent when any demolition is done. The contractor should prudently have pointed out any likely areas of extra work to the insurer. In addition the architect will certify the interim payments for the reinstatement work each month. These sums will be included in the normal series of certificates but will not be subject to retention as this has already accrued under the initial payments for the damaged work. It will therefore be necessary for the employer to withdraw equivalent sums from the joint account to meet these commitments. An inclusion of this kind is shown in Case A in Chapter 24.

The quantity surveyor's position
So far as any direct agreements with the insurer go, the quantity surveyor should maintain an interested but neutral rôle, unless he has been appointed as loss assessor. Under the contract the contractor is to receive the whole of the insurance monies other than the professional fee allowance; while he receives no less, nor does he receive any more. The agreement of the monies is something outside the contract and therefore not something into which the client or his advisers will normally be able to intrude.

In the normal circumstances of an entirely independent assessor acting, the surveyor will mainly be interested in reconciling the settlement with the physical work to be reinstated, so that he may calculate the interim payments on a reasonable basis. There are two sides to this. He should first form an opinion, as quickly as possible, of the stage that the works had reached in the various parts that have been damaged; this may not be easy since some sections may have collapsed or have been carried away. The insurance monies will be paid out so that they are completely cleared when this line of demarcation is reached in the various parts. A knowledge of this demarcation will also help in deciding when further payments from the employer himself have become due, although this decision should be reached by the usual processes of valuation and should not depend upon a line drawn perhaps arbitrarily for another purpose. The previous payments may themselves have been arbitrary in certain respects and their timing is unlikely to coincide precisely with the damage. The extent of undamaged work from which the reinstatement is to begin will, in the nature of things, be somewhat easier to determine.

Secondly, the surveyor will need an analysis of the settled sum, so far as this is available. While the settlement has been the contractor's concern, the surveyor is justified in asking for such detail of its apportionment as there may be; an analysis supplied by the assessor is desirable. Wherever it may come from, it should be used with an awareness that its method of calculation may mean some adjustment before it can be applied to its new purpose. If all else fails, the surveyor must prepare his own analysis from the estimate adjusted in proportion to the level of the settlement. The extent of removal of debris is an item not to be underestimated in such cases.

Where there is a definite settlement in advance of the work being done, there is only one point at which any problem of the employer paying twice for the same effort arises. This is in respect of fluctuations calculated on the traditional basis and distinct records must be kept of the expenditure on the reinstatement work, so that any fluctuations may be deducted from the main account. Otherwise the account is settled as though no damage had occurred at all. Nothing can be recovered under the insurance for extra fluctuations due to delayed completion. Where a settlement is reached by an agreement to pay the cost of reinstatement, the position will be essentially the same for the normal measured or other lump sum contract to which clause 22A applies. If a clause similar to this clause is incorporated into a prime cost contract there will be far more than just the fluctuations at stake if there is any overlap. The JCT Fixed Fee Form does not make such a clause a standard provision, possibly for this reason.

The costs eligible to be covered in the contractor's insurance claim are set out under the next heading, but it must be noted that he may have other expenses involved in the reinstatement and not specified in the clause. If any of these are of relevance they may need attention over any overlap.

The contractor's position

The contractor will find himself in immediate danger of answering to all manner of people in a case of extensive damage but he still remains in his basic contractual position and is thus responsible for producing the works and taking any instructions under the contract from the architect alone. Any parts of the works that are undamaged he must still proceed with, unless the damage has affected the practicability of access or some other controlling factor. Any damaged part that he allows to deteriorate further he will still have to replace and he should take steps to avoid its decline, so that the insurer does not demur over the amounts of settlement. Should the insurer require the contractor to take any steps in this direction, he should do so without an architect's instruction, since these items will not fall under the contract for settlement. Where the building inspector or district surveyor requires any demolition to take place, the contractor should seek that this be routed through the architect. An insurer under clause 22A will usually take an even view of any such work, whatever its origin. Where the employer is responsible for taking out the insurance, the reinstatement is deemed to be a variation; there will seldom be room for serious difference between the inspector or surveyor and the architect. The contractor should nevertheless operate the provisions of clause 6.1 to avoid any doubt.

The contractor will be even more interested than the surveyor in the extent of the work to be replaced, since he has to agree the claim amount with the insurance company. If a lump sum is arrived at, any uncertainties that depend on demolition still to be carried out should be specified as excluded from the settlement so far, so that the door is not closed to a further amount if this is justified. The basis of the agreement reached need bear no direct relation to the rates in the contract, which are for work carried out in a distinct rhythm from that of remedial work. The contract itself considers insurance from the point of view of protecting the employer ultimately, and therefore restricts not only the list of risks to those most likely to affect the installed work, but also restricts the scope to that work and to the unfixed materials intended for the works. In so doing it requires the direct reinstatement only to be covered.

Beyond these aspects, the contractor should reasonably have covered by insurance a number of other points. These will include his materials off site, whether these have been paid for by the employer or not, his plant and other equipment on site, and also the costs of delay as such. While a clause 22A insurance covers the immediate supervision and other overheads associated with the reinstatement, it will not cover the contractor against the disruption of his programme and the standing time and other costs incurred. A normal all-risks policy will cover a number of these items and the rest may be covered by supplementary premiums.

In relation to the contract, the contractor will wish to obtain an extension of time for any serious incident. This must be granted in any case where a delay has been caused, even though the contractor may have been negligent in what led up to the damage. From what has been said, and from the silence of clause 26.2 it will be clear that no claim can arise under that head. A determination notice from the contractor is feasible, as has been mentioned already. However, it is usually the case that all those participating in the various stages of settlement and reinstatement pursue their functions with the aim of tidying matters up as conveniently as possible. An unduly harsh or niggling line on the part of anyone does not help the immediate problem to be resolved and does little for the goodwill of future dealings that is still relevant today.

CHAPTER 22

Insolvency of the contractor and determination of his employment

While the subject matter of the last two chapters has had a sombre aura about it, that of the present chapter is in some ways even more depressing. It brings delay and loss to the client—not all of which he can recover—and it brings shipwreck to the contractor and perhaps to those who are in his wake. It brings extra work to the professional team and, while work is usually welcome, this particular work is something that nobody enjoys or even profits by.

It is necessary to explain the particular limits set for this chapter. It is limited to the contractor's collapse, because this produces a more complex situation in settlement than does the collapse of the employer. In the latter case the contract is wound up by paying the contractor for what he has done and the the loss that he has suffered, which is procedurally simple though complex in calculation (see *BCPG*, Chapter 7). The scope of this chapter is further limited to those cases where the contractor leaves the site because of his insolvency. Some reference is also made to the reinstatement of the contractor as another possibility; if this is done, and work proceeds to finality, settlement will continue as usual. He may, alternatively, go either before or after insolvency because he is in default; here the resulting actions will be similar to a number of those discussed in this chapter but they will not be complicated by the special issues connected with insolvency only. Actions taken on default can therefore be picked out fairly easily from the matters mentioned in this chapter without their being underlined in passing.

Finally, the term 'insolvency' is here used as an umbrella to cover bankruptcy or liquidation and what accompanies them. Case B in Chapter 25 follows through a set of financial workings that are typical of what happens when these events occur and it gives form to a number of the points discussed below.

The fact of determination

The standard forms of contract all incorporate provisions aimed to regulate, as far as is ever possible, events at or about the contractor's insolvency, but a number of legal problems affecting validity hang over this whole area. The JCT form employs the term 'determination of the contractor's employment under the contract' to emphasise that the contract still remains in force to cover the outstanding issues. The other main forms use different terms but reach much the same position nevertheless. Only the JCT procedure therefore is considered specifically in what follows—and in the context of bills of quantities at that (see *BCPG*, Chapter 7 extensively in what follows).

Where the JCT form is on its own is in its attempt to make determination automatic upon the contractor's insolvency becoming certain. This is an attempt that may be regarded kindly, since the actual moment of insolvency may not be clear until after the event is well passed. However, it is undermined by some of the legal problems referred to above and it is therefore as well for the client to obtain agreement by the liquidator or the like that he accepts determination as effective. If he does not do so it would seem that he is backed by statute law and may choose within 28 days to carry on with the contract. This may be a welcome move from the client's point of view, since there will be continuity; the drawback is that the liquidator (as he will be called hereafter) may well exercise his right to disclaim the contract at any later stage—perhaps when the more fruitful work has been performed, or when some retention has been released if a partial possession situation arises.

Usually, however, once work is fairly well advanced the liquidator is not likely to disclaim if he can avoid it. A relatively small amount of work is going to release the retention, whether it be the whole or an outstanding proportion. If he does disclaim at a late stage the client will have things at their most favourable, since he will have the funds in hand to meet the more expensive costs of completion. Whether they will also cover his loss due to delay is another matter. There is some legal doubt as to whether clause 27.4 will still be valid after a disclaimer or whether it will be necessary to prove loss in the normal way. Most clients will prefer, on all counts, that the work is not interrupted, if at all possible.

If work is in a relatively early stage when insolvency occurs, it is the client who will have the stronger incentive to seek a reinstatement. He will have less money in hand, the prospect of having to pay for a much greater amount of outstanding work at higher prices, and the likelihood of a greater delay before his building is available. Since he is unlikely to recoup all of his loss, he may well be excused his depression. The liquidator in this situation will face more work to arrange for if he elects to continue, with a proportionately smaller return. He may well be strongly inclined to accept the determination; he certainly cannot be forced to continue.

If the client finds himself faced with a contractor who is continuing, with support of his liquidator, he may still have a way out if he wishes to be free of a bad performer. The right to determine on grounds of the contractor's default in performance still remains, since the liquidator, in proceeding, is bound by all the terms of the contract, including the requirement to make proper progress. If such progress does not occur, the client may well acquire a valid right to determine independently of the insolvency. In all these situations there will be a number of factors to take into account in making a decision, and not all will point in the same direction.

The immediate position

Assuming that a determination of the contractor's employment comes about, it is the position in which the client finds himself that should occupy first place in the attention of his advisers. A great measure of sympathy may or may not be felt towards the contractor or, more likely, towards his suppliers and sub-contractors, but this must remain as a sentiment when giving advice or acting. The primary aim must be to help the client to recover the situation so far as is at all possible. His immediate position and his overall actions may be considered first before the distinctive rôle of the quantity surveyor is outlined.

MATTERS WITHIN THE CLIENT'S CONTROL

Until the contractor's legal insolvency is a fact, the client is in a rather frustrating position. He must continue to make any payments that fall due under interim certificates, knowing that he may not see the benefits of his money in the ultimate settlement. He cannot enter on the site to affect directly what is happening there. Sub-contractors may be doing what they can to protect themselves against the coming storm and the architect will need to control any attempts to remove materials from site, where this is contravening the provisions of clause 16.1.

At the time of determination the client will have a number of items immediately available to him or within his control. The most important of these are:

(a) The works themselves, so far as they have proceeded: these he undoubtedly owns.

(b) Materials on site for which he has paid on an interim certificate: these too he will own unless the contractor's title is defective under the wording of a particular sub-contract or contract of sale.

(c) Materials off-site for which he has similarly paid: these he will also own, provided that the provisions of clause 30.3 have been properly followed or the corresponding contract of purchase has been used in Scotland (see BCPG, Chapters 11 and 26).

(d) Materials on site for which he has not paid: the contract gives a right to retain these on site if required and to use them in completing the works. No payment is to be made directly, since they will reduce the total cost to the client of achieving completion; they will also reduce the final indebtedness of the contractor that almost inevitably arises and thus are, in effect, paid for at that stage.

(e) The contractor's own directly owned plant and other temporary equipment and structures on site: these may be used freely to secure completion, either by the client or by a new contractor on his behalf. Hired plant, and that of sub-contractors, cannot be so retained. No plant at all passes permanently into the ownership of the client.

(f) Retention: this is retained by the client and is one of the main buffers that he has with which to take up the loss that he faces. That part of the retention fund that is held in respect of nominated sub-contractors still constitutes a trust fund (see *BCPG*, Chapter 16) and must therefore be conserved for payment to those firms in due course, the time depending on the terms of any new agreements reached with them.

(g) Any balance existing as a result of underpayment in interim certificates: this will have the same effect as an additional amount of retention. Its very existence may not ever be certainly known, as is discussed in dealing with the quantity surveyor's action; if it were known it might, in any case, turn out to be of negative value.

A number of other stray items may also be thrown up, such as payments for water that have been made on a lump sum basis for the whole of the works, as the practice is in many areas. If there was a bond in the contract requirements to secure due performance then this will lie in the background as a reassurance to the client that any overall loss that he may otherwise finally face will be cushioned to the limit of the bond. Initially, however, the bond is no source of funds or means to proceed.

ACTIONS BY THE CLIENT

The term 'client' is here used to include also all those who may be acting on his behalf, particularly the architect (the salient actions that will usually fall to the quantity surveyor are dealt with separately below). What everyone should do during the period before insolvency is established to avoid creating an atmosphere of panic that may well hasten the contractor's collapse, or even precipitate it when it might otherwise have been avoided. Spreading rumours and the like may be damaging to the originator as well as the subject.

There are several actions which the client will perform as soon as possible upon the contractor's insolvency. He should confirm with the liquidator that he is accepting an automatic determination; this should be done in writing, of course, and early enough to allow for any moves over sub-contracts to be made. Such confirmation also resolves any doubt hanging over affairs and allows further matters to be put in hand. These may include seeking assignments of agreements that did not include an automatic determination provision of their own. Even more urgent may be the stopping of an interim payment that is already on the way through the banks; there is no obligation to pay after determination, even though a certificate has been issued, and the client should not do so if he can avoid it. Obviously, it must be fairly well established that there will be no attempt to proceed before this is done, although some would urge a policy of 'shoot first and ask questions afterwards'.

The site itself needs attention. A proper arrangement for security may mean erecting additional hoardings to those of the contractor, locking sheds and providing a 24-hour guard. No one should be allowed on site other than direct agents of the client or the liquidator, and initially nothing should be removed from site at all. Inventories must then be prepared of all materials, plant and other movable or temporary goods on site. So soon as the situation has been thus stabilised, attention may be turned to the pressing horde at the gates seeking all manner of property. The plant of hirers and sub-contractors may be let off site when its status has been cleared, and provided that no fresh agreements involving its continuing use are in prospect. If any of the contractor's own plant is unwanted it should be ordered off by the architect. If the contractor does not deal with this within a reasonable time the plant may be sold and the net proceeds held for the contractor.

Apart from action on site, there will also be a need for a fresh insurance of the works to replace the contractor's insurance under clause 22A if this has been in force. Such a policy could well be taken out with the fresh contract in mind, since this will require the works carried out to date to be treated as an existing structure. The new contract insurance clause would therefore be clause 22C.

ACTIONS BY THE QUANTITY SURVEYOR
The first action by the surveyor will undoubtedly be to check through his valuation details to see whether there are any danger signs of over-payment. He should have been paying attention to matters such as unpaid firms and the running down of the tempo of the job that almost invariably precede insolvency. He will have been in the difficult position of having two conflicting sets of obligations. On the one hand he will have been concerned to protect his client by avoiding any overpayment, but against this he will have weighed the desire not to keep back the contractor's due entitlement and thus push him over the brink that much sooner and perhaps at all, to the detriment of all concerned. Particular care in preparing valuations is necessary at the stage when they are likely to have far more than interim significance.

No doubt he will be asked to give an assessment of the state of the finances and a close viewing of his valuation figures will help here. In view of the other matters, many of them imponderable at this stage, that will need to be given a financial tag, excessive accuracy need not be sought. A reasonably good check on his figures against the detailed site situation is useful in case figures are needed for some evidential purposes later in an unforeseen dispute. It is not necessary for settlement purposes to carry out a precise full remeasurement valued at the contractor's rates; he is not paid on this basis in the final settlement. The only point of any such measurement will be if it helps in the preparation of a bill of quantities for the completion contract, as is mentioned later.

A review of firms claiming that they have not been paid will also be needed. There is no legal obligation to pay any such firm, nominated or otherwise, and the contract does not permit such payments. Any payment made is voluntary and may be risky, although a moral pressure in favour of a nominated firm may be felt. More strongly, there may be a business pressure in that the firm may not be willing to return to site under a new contract unless it secures at least part of its past money. Anything so paid by the client will reduce the margin that he has available to him in the final balancing, when he is likely to need every penny. There is also some doubt whether in law the client may not find himself liable to pay such amounts again to the liquidator; the latter should be implicated in any such agreement to pay, to avoid this risk. If there is a bondsman, he too should be consulted. Only retention accruing before determination may and must be paid over to nominated firms, as and when it becomes due.

A number of the inventories needed during the transitional period will come the way of the surveyor and these should be prepared accurately, since various persons may have need of the information in them. This is especially true of any plant inventories. While the client is not responsible for fair wear and tear on plant, he is responsible for any losses of plant unless the contractor fails to remove it when required.

Completing the work

ARRANGING A FRESH CONTRACT OR CONTRACTS

Much will depend on the stage and state of the works as to how their completion is to be dealt with. If a fairly large proportion of the work remains to be done in terms of main contractor's work, then a measured contract is appropriate and a bill of firm quantities is desirable. If the amount of work carried out under the original contract is fairly small, then the bill may be prepared by measuring the work done and subtracting it from the total original bill, adjusted by any variations. Otherwise it may be simpler to start from scratch. A firm bill is in any case going to lead to delay in restarting the job and this may not be acceptable to the client. Delay may be reduced in two ways here. One is to settle for approximate quantities and remeasure; the other is to place an initial contract for a defined amount of work on a prime cost basis, to be carried out while the firm bill is being prepared.

This latter arrangement has a virtue of its own, in that it allows certain preparatory work to be done that would be dealt with on a day-work basis anyway. This will include the completing of half-done jobs, the remedying of work discovered to be defective, and the moving of materials that are in unsuitable positions. Even if these jobs are done in advance, there should still be a sizeable provisional allowance in the completion contract to cover more of such work. Good supervision on the part of the original contractor and a methodical approach are hardly likely to have been his strong points (if they had been, it is highly possible that he would never have got into difficulties in the first place) and the full extent of his errors of omission and commission may not come to light for some time after the new contractor has moved onto the site.

If work is near completion at determination, and what is left is largely piecemeal it may be better to arrange a contract on a prime cost basis for the whole. Sections of the work may however be self-contained and lend themselves to distinct lump sum contracts of their own. This may suit an unstarted car park for instance.

However the work is arranged, there will be a number of provisions to be included in the contract documents that arise from the peculiar nature of the work. These will relate particularly to procedures for making assignments of agreements from the client, who may have taken over such items from the original contractor, and also to materials and plant on site. The contractor coming in must agree any inventory of items that he takes over immediately upon entry in bulk; he must also be fully aware of any procedure that he will have to operate for withdrawing items from any site store controlled by the client. The bill must be prepared in such a way as to allow either fixing rates or credit terms for these items, as is the more suitable.

PROCEEDING TO COMPLETION

The new contract should be settled in the same way as any other contract of its type and presents no fresh problems. The worst that could happen would be for the new contractor to become insolvent and for two layers of accounts to superimpose themselves on one another. Any surveyor so encumbered should consider early retirement! While agreement of the final account is going on, however, a careful note will be made on the reason for all items that are embodied in it; some of these may have a special significance in settling the old contractor's account later, as is discussed below.

The other action that may need the quantity surveyor's attention is the clearance of the original contractor's plant and other property from the site. The contractor is required to do this when the architect so asks, but if he fails to do so then the architect may arrange to sell off the plant; the original contractor is entitled to the proceeds, after any expenses of sale have been deducted. In this second case he would lose his right to recompense for any loss or damage. In each case it may be required of the surveyor that he reconcile the disposals with any inventory that he prepared at determination.

Settlement with the original contractor
While events have been grinding or jerking towards the physical completion of the works, the contractor's liquidator will have been waiting in the resigned manner of such persons for the outcome. At the stage of entry into any fresh commitment by the client he may well have been advised of the proposed basis and terms, since he is able to query any extravagant measures that are taken. Provided that normal commercial prudence has been exercised, he will not now be able to complain. This means that a sensible balance has been struck between time and money in bringing the works to completion, not necessarily that every penny has been saved.

Several elements enter into the final reckoning and these are discussed in principle here. The first three represent the client's total actual expenditure on the works as intended and on certain completely extra items, while the last is the corresponding amount that he should have paid. The difference is the debt due from one party to the other; it needs little imagination to discern its usual direction.

THE LOSS AND EXPENDITURE FINALLY INCURRED
The client's direct loss — The main source of loss to the client will be the late completion of his building. If the original contract had gone to completion the measure of this loss would have been liquidated damages; these strictly lapse with the determination but may still be adopted by the parties as the most convenient assessment available. If, however, some significant change has occurred in the overall circumstances a new level of loss may have supervened and may be urged by either party; this would not have been possible with liquidated damages in a continuing contract. Into the direct calculation of this sum the surveyor will possibly not enter but he may need to give advice on the period to which it applies.

In general the client will be able to charge the loss for the whole period of delay between the time when the building should have been completed and the time when it was actually finished. In both cases he must allow into the calculations the extensions of time granted to the contractors, on the assumption that the full amount would have been granted to the original contractor if he had continued. If the completion contractor was made liable for liquidated damages the period concerned should be deducted from the eventual period and not, therefore, charged against the original contractor. There will have been delay in restarting work after determination and this is chargeable as part of the delay, unless there was some unusual cause for any part of it. The completion contract may have had a contract period that gave a faster or slower rate of progress; the contract price has reflected the difference in cost and is incorporated as it stands in the settlement and so the client's loss must be adjusted for this difference also.

The client's expenses in completion — The sum of the various accounts for completion of the works will be produced and a number of ancillary items will arise. These may include temporary protection and other measures while the works were standing between contracts, the additional insurances that the client has borne then and through the new contract, and the fees that have been paid for extra services in placing and managing the additional contracts. Fees on the hypothetical account mentioned below will also come into this last reckoning; in addition, there may have been payments of retention money to nominated sub-contractors on sums paid net of retention through the original contractor.

Any direct payments made to firms not paid by the original contractor will be included with other accounts at this stage and it is now that the need to justify these payments may arise.

The client's payments to the original contractor — These are simply the totals of interim certificates as honoured by the client, with the addition of any sums paid direct before determination and deducted from the amounts due at that time. Payments made direct after determination must not be deducted, since they have been made twice—once under this head and once under the preceding.

THE EXPENDITURE THAT SHOULD HAVE BEEN INCURRED

The more difficult side of the balance, from the surveyor's point of view, is in arriving at the quite hypothetical story of what should have happened. In outline it is necessary to prepare a final account for the works as actually carried out, based upon the original contract bills and allowing for all adjustments to the contract sum that would have arisen. This account is then priced on the original basis also.

For most items this is essentially an exercise in translation. Measured variations and dayworks will be repriced at the original rates, if carried out in the new contract. A watchful eye must be kept for any daywork items that might have been measured but for the interruption factor that has been introduced. Where work has been performed under an entirely prime cost arrangement, this side of things will tend to reverse itself and the surveyor will need to look for elements of the account that should still fairly be dealt with as daywork. This requires a good deal of discretion, since the time expended may be particularly troublesome to extract from the account, where it may well not have been segregated unless the necessity was foreseen at the time the work was done. This is more easily said in a book than it is done in practice and the best possible often has to be done in awkward circumstances.

Loss and expense items incurred in the completion contract can only be assumed to have been of the same magnitude as they would have been under the original and they too must be repriced if necessary. In all parts such as these, which may not be closely related to contract rates, the main factor to consider may be the difference in basic price levels not taken up in the fluctuations of the completion contract. So far as fluctuations themselves are concerned, there will be two facets to any repricing; one will be the difference in basic price levels underlying the measured rates, while the other will be any excess fluctuation increases due to the late timing of the work. Since liquidated damages are not being levied as such, these increases may fairly be deduced; this is distinct from the position set out for normal fluctuation calculations in Chapter 19. Where one of the two contracts was not subject to fluctuations, it will be necessary to assess the allowance in the rates for offering a fixed price, so that comparability is reached here also. These adjustments are easier to deal with under the formula system, since the index numbers avoid making a lot of assumptions.

Prime cost sum expenditure should be reviewed along similar lines. The mere fact of a later acceptance of a quotation does not lead to an adjustment downwards, since an earlier acceptance might have been followed by a larger element of fluctuations, but the date of carrying out the work may again have a bearing. Where a nominated firm was under contract to both contractors, it may be necessary to weed out any surcharge that the second quotation carried to reimburse the firm for the disruption suffered and, perhaps, to persuade it to resume the work.

The foregoing represent what are broadly accounting differences. In addition there are likely to be items chargeable by the second contractor which would have been carried by the first contractor without them becoming extras on the contract sum. These will include corrective work, either to make good defective construction or to reorganise an untidy site and perhaps dispose of materials that have deteriorated and equipment that is unsuitable. It sometimes happens, especially with regard to groundworks, that the most straightforward way of overcoming a fault is to redesign some part of the works around it; this will result in a variation that must be neutralised in the present exercise. This can become a speculative matter if a further genuine variation is then superimposed on this basic change. It may have occurred anyway, but it may just be a case of taking advantage of circumstances as they have fallen out. There is no rule for such complications, they must be considered on whatever merits they may suggest for themselves.

The resulting account should be drawn up in as near a form as is possible in the circumstances to what would have been produced if it had been the actual account for the work. This may be modified if it is helpful for comparison between the various documents to show things in a different format. The detail of sections may also be such as to give information on how calculations were made; this is useful for the surveyor and his readers when apportionments or overall percentage adjustments are employed. It is far better to disclose what has actually been done on the face of the account than to attempt to produce a make-believe appearance.

The figures in Chapter 25 do not show a typical layout, since they are concerned with underlying principles only, but the foregoing comments may explain something of the approach in what is produced. The same figures also show how all the figures involved in the final settlement come together and may be left to speak for themselves.

PART 5

TWO CASE STUDIES

CHAPTER 23

Some preliminary comments on the case studies

The two cases given in the following chapters are not intentionally portrayals of actual contracts; any resemblance that they may suggest is therefore to be regarded in the way that it usually is in the fiction realm of literature. At the same time the situations that develop are not untypical of many that crop up from time to time. The first case proceeds entirely by presenting the interim valuations that are made during progress and the second case follows the same pattern for parts of its course. This method has been used partly because this part of a surveyor's work is not so closely related to the detail of taking off quantities and other processes that fall into the area of technical operations that the present work does not aim to cover and which is covered by a substantial bibliography already. This distinction from technical operations is particularly true of the handling of preliminaries, for which the amount of space used may otherwise suggest an undue predilection. The method has also been used because it is in interim payments and their calculation that the contract conditions come most obviously into play.

In both cases the conditions assumed are the standard JCT form, in the 1980 private edition with quantities, although little would be affected by the substitution of one of the other editions if enough detail of the pricing were obtained in the case of the editions without quantities. The retention provisions have been amended for Case A where a limit has been put on the total fund. The change allows one or two complications to be introduced for illustrative purposes: these can easily be ignored for anyone wishing to follow the standard pattern, but would be troublesome for the reader to introduce to the printed text. It is illuminating to increase the values for retention at various stages in the second example and to note what an effect this has on the amount of funding of the works that has to be undertaken. This is a major aspect of a contractor's operations that perhaps receives less consideration from the surveyor than it might, although very properly he will always consider it alongside maintaining the correct safeguards for the employer. The point might be explored further in relation to the contractor's cash flow arrangements (see also Chapter 4).

A drastic simplification of many details has been inevitable; this is necessary to keep the whole within bounds and to isolate the key elements, so that the figures may be easily followed. The presentation of the calculations for preliminaries is especially affected here; while a reasonable method of allocation has been adopted, it has been the policy to keep the figures rounded and to assume that any delays in programme have a simple effect so far as retiming the payment of preliminaries goes. This same simple effect has been assumed in dealing with any extensions of time and any payments for loss and expense. A system of 'instant' agreement has been operated that would be the pride of the industry if it were found to be operating in practice. What applies to preliminaries in this respect is also helpful over the other parts of the value of work.

In some instances the same figures have been used to assist in tracing a connection, although they would not correspond so closely in practice. Thus an apportionment of preliminaries has been made in Case A for valuation purposes and it has been stated that the results might well not apply for any other purposes. Nevertheless, the same figures have been retained for example purposes later in the history. In a similar way, interim valuation figures have been used as the approximate estimate amounts for the purpose of sectional completion and partial possession payments and damages, without the refinements that would actually be applied.

Some violence has also been done to the proper order in which many of the figures given would appear if they had been given in their normal context of a fully detailed valuation or final account. Again, the desire to present salient features, rather than to give a step by step example of how to draw up the documents concerned, must be pleaded. In many places approximation of figures (often to the nearest £50 or £100) has taken place, again for simplicity of reading.

CHAPTER 24

Case A: A contract completed

This case study might be sub-titled 'per ardua ad astra', as the project makes it way through a number of troubles without succumbing. There are a number of failings by the employer and the architect in mid-stream and a closing lapse by the contractor. In addition there are a few misfortunes leading to results that are shared between the parties. But on the whole a reasonably happy landfall is made at the end of the day. Some of the more important elements in the saga are the following:

(a) The programme shows both partial possession and sectional completion occurring. There are also delays in the programme, with the blame to be laid at more than one door.

(b) The retention percentage starts higher than the JCT recommended level for the size of the contract sum, but then has a maximum at the recommended level. Retention thus builds up more rapidly during the earlier stages, but then no more is held on later work. This is more severe throughout than the JCT pattern. It also leads to several quirks when early final payments and completions occur.

(c) Several sums fall to be paid gross during progress, that is without deduction of retention. These include claim items, and some insurance items but not others.

(d) The preliminaries have been priced in reasonable detail and so a fairly detailed analysis of them for payments is undertaken.

Firstly the key points of the contract are set down and some analytical work is done. Then some valuations for crucial months are outlined and commented upon.

Some aspects of the contract

THE CONTRACT BILLS
Much of the content of the contract bills is of no direct interest for present purposes.

The following are the main totals:

		£	£
Bill 1: Preliminaries			320,000
Bill 2: Preambles			—
Bill 3: Office block	PC sums	280,000	
	Measured	600,000	880,000
Bill 4: Stores extension	PC sums	416,000	
	Provisional	64,000	
	Measured	720,000	1,200,000
Bill 5: Process building	PC sums	1,040,000	
	Measured	880,000	1,920,000
Bill 6: External works (all measured)			240,000
Contingencies			240,000
Water, insurance and errors			32,000
Contract sum and total of bills			£4,832,000

THE CONTRACT CONDITIONS AND APPENDIX

Two choices within the conditions are important for what follows. First, 'full' fluctuations apply and clause 39 has been included. Second, so far as insurance is concerned, this is being taken in the main by the contractor under clause 22A but the employer is responsible for the stores extension under clause 22C.

In the appendix the key insertions agreed are:

(a) Date for possession		Date X
(b) Date for completion:	Office block	Date X plus 15 months
	Remainder	Date X plus 20 months
(c) Liquidated and	Office block	£1,500 per week
ascertained damages	Remainder	£8,000 per week
(d) Prime cost sums for which the contractor desires to tender		Ceiling linings
(e) Retention percentage		5%
(f) Limit of retention		3% of contract sum

Three comments are needed. Item (b) means that the sectional completion supplement has been used in the contract and overlays the provisions of clause 18 in places, without cancelling them. Following on this, item (c) gives fixed proportions to the damages which are not those which would be produced by the formula in clause 18.1.5. Item (e) gives the retention level which would apply as a flat rate if it were not for the special provision.

Wording has been introduced into the contract to keep retention on nominated sub-contractors in proportion to these levels—as the JCT form once did itself.

NOMINATIONS FOR PRIME COST WORK
Only a few of the large number of PC sums need to be mentioned here.

(a) Nominated supplier for reconstructed stone (office block)

(b) Nominated sub-contractors for:
  (i)   Ventilation installation for office block
  (ii)   Ditto for process building
  (iii) Concrete piling for stores extension
  (iv)   Steel flooring for stores extension

All of these were put to the contractor with quotations that allowed 2½% cash discount. The contractor objected to this in the case of the first item and the architect agreed to an addition being made in the final account to cover the difference. This saved having to obtain a revised and increased quotation from the firm. If however the contractor had accepted the nomination without question he would have lost his right to the discount. The last item, for steel flooring, was not covered by a PC sum in the bills but arose out of a provisional sum, as an alternative acceptable way of a nomination being made.

In addition to these nominations, the architect accepted the contractor's own tender for the ceiling linings as noted in the appendix. The contractor had been instructed when tendering to include an allowance in lieu of cash discount, so that his tender was directly comparable with those of other firms. For the same reason, he was also told that profit and attendance would be added to his account for the work, as it had been to the PC sum in the bills.

In what follows, it has been assumed that the accounts for nominated work total up to the same amount for each bill as the amount originally given by the PC sums, except for the net addition for fluctuations.

Some preparatory analysis
THE PRELIMINARIES BILL
Several items were priced individually in Bill 1: Preliminaries as follows:

	£
Setting out the works	16,000
Programme preparation	8,000
Foreman in charge	64,000
Transport	40,000
Plant (endorsed as including cranes and hoists)	136,000
Site accommodation	48,000
Clearing at completion	8,000
Bill total	£320,000

During the examination of the tender, the quantity surveyor agreed with the contractor an analysis of these items and of the general summary additions, which it was stated would apply for valuation purposes. If an adjustment of preliminaries was necessary for final account purposes, then this present analysis would not be binding and a fresh one would be made in any greater detail that was appropriate. First the items were considered separately:

(a)   *Setting out:* This would occur mainly in the first three months and was spread over them. A smaller proportion of setting out would occur in the closing stages for the external works; this was ignored, since it would be compensated for by the way in which the next group of items was spread throughout the whole period.

(b)   *Programme, foreman and transport:* These were spread evenly throughout the twenty months, despite the early handing over of the office block. The reasoning was:

(i)   Programme preparation would be heavy initially and then be largely a matter of reviewing and the detailed integration of nominated firms. This gave an early hump in the effort.

(ii)   Apart from the agent and others engaged from the early days, the item for foreman covered also staff who would build up in the middle term of the project and then disperse slowly in the latter stages. This gave an 'end heavy' weighting to this expenditure.

(iii)   Transport would rise with the contractor's own direct effort on site and would have a distinct hump in the middle period.

These items were considered to even out within the limits of accuracy of payments on account, bearing in mind also the amount of setting out paid for early. If anything, there was a slight tendency to pay late.

(c)   *Plant and accommodation:* The two items specified were hired; the rest were owned by the contractor. The spread of payment thus varied in emphasis:

	Plant hired £	Plant owned £	Accommodation £
Erection	4,000	48,000	24,000
Running	56,000	20,000	20,000
Removal	4,000	4,000	4,000
	£64,000	£72,000	£48,000

These figures were then allocated early and late for the first and last elements and spread evenly for the others.

(d)   *Clearing*: This was allocated on a broad proportion basis between the two stages of completion.

(e) *General summary additions for water, etc:* These were fairly modest in total and were taken at two thirds per cent of all work, whether the contractor's own or not. Since his work tended to occur early this could be held to be favourable to the employer. Water charges, if payable in advance, could even have been paid as lump sums.

The sums spread evenly over the period were thus the following:

	£
(b) Programme etc.	112,000
(c) Plant hired	56,000
(c) Plant owned	20,000
(c) Accommodation	20,000
Total for 20 months	£208,000
Amount per month	£10,400

The other items fell to be allocated to the early and late stages of the project in accordance with this table:

Month	Setting out £	Plant £	Accommodation £	Clearing £	Total £
1	8,000	8,000	8,000	—	24,000
2	4,000	16,000	8,000	—	28,000
3	4,000	20,000	8,000	—	32,000
4	—	8,000	—	—	8,000
15	—	800	—	1,600	2,400
18	—	1,600	1,600	—	3,200
20	—	5,600	2,400	6,400	14,400
	£16,000	£60,000	£28,000	£8,000	£112,000

THE MEASURED BILLS

Bills 3 to 6 were analysed in conjunction with the contractor to enable the employer to deal with his own cash flow arrangements. The results are set out in Appendix 1 to this chapter and include the sums actually paid for the first six months of work. After that, there was a slip in the programme, followed by others, and the payments actually made were the results of the figures set out in Appendix 2, which show the total sum spread over 24 months instead of the 20 months originally allowed for.

With this investigation of the work done on the finances before physical work began, it is now possible to pass to the happenings.

Position at the end of selected months
MONTH 2
Everything has proceeded to plan and the resultant valuation at the end of the month is prepared accordingly. It is given as representing an unsullied state of affairs. Fees do not suffer retention, it may be noted.

		£	£
Preliminaries:	Setting out	12,000	
	Plant	24,000	
	Accommodation	16,000	
	Regular items (10,400 × 2)	20,800	72,800
Bills 3 to 6			232,000
			304,800
	Water etc. 2/3 %, say		2,000
	Total work		306,800
	Materials on site		76,000
			382,800
	Less retention 5%, say		19,100
			363,700
	Local authority's fees		2,400
			366,100
	Less previous payment		159,600
	Valuation 2		£206,500

MONTH 8
The scheduled payment was made at the end of Month 6 but early in the following month a delay was occasioned. The cause of this was an explosion in a workshed standing between the office block and the stores extension. As a result there was a substantial disruption due to actual damage and also to a slight delay in securing acceptance of the insurance claim on the office block, where the contractor had insured. The contractor was granted a one month extension of time over the whole contract. Only £9,400 fell to be paid on Month 7; this was paid, since the contract has no provision for any minimum payment to justify a certificate. The programme has now regained momentum, just one month behind, and part of the repair work has been carried out in addition.

The insurance settlements are several:

		£
(a) Contractor's policy under clause 22A	Office block	32,000
	50% of materials	
	in shed	12,000
		£44,000
(b) Contractor's policy outside contract:	Workshed	£16,000
(c) Employer's policy under clause 22C	Stores extension	16,000
	Existing stores	4,000
	50% of materials	
	in shed	12,000
		£32,000

The contractor has agreed to perform the repair work in the existing building at contract rates, as it is quite minor. This he need not have done at all as a matter of contract responsibility. It will be seen that there was an overlap of two contract insurances in this case and that the two companies resolved it by an equal split. There is always a risk of complication in dividing any risk in this way and it was only done here because the buildings stood quite separately. It is better to let the contractor take his normal responsibility, where this is possible, on as much work as may be. Neither the contractor nor the employer is covered by his insurances for the cost of the delay that has been suffered, although they could have had such a cover in addition to the contractual requirements. Amongst other things, the contractor receives no extra preliminaries and the employer loses liquidated damages and has to pay any extra fluctuation increases.

The contractor has, however, requested payment on account for materials stored off-site as a result of the disorganisation on site. He has also asked to be paid in the final account for the cost of this storage. Both these requests have been declined as being in respect of consequences that he must bear—although a concession on the former count would not have been unreasonable.

The valuation reflects the foregoing position and some comments on its detail follow it.

		£	£
Preliminaries:	Early items	92,000	
	Regular items (10,400 × 7)	72,800	164,800
Bills 3 to 6			1,136,000
Repair work, clause 22C			20,000
			1,320,800
	Water etc. 2/3%, say		8,800
Total work			1,329,600
	Materials on site		184,000
			1,513,600
	Less retention 5%		75,700
			1,437,900
Local authority's fees		2,400	
Repair work, clause 22A		24,000	26,400
			1,464,300
	Less previous payments		1,233,700
		Valuation 8	£230,600

Since the programme has suffered a clean break and has resumed it has been possible to put back the payment of preliminaries by just one month, rather than to recalculate them as spread out over a longer period.

The divisions of the repair work have been treated differently as far as retention is concerned. Under clause 22A, the monies are payable at the same time as normal interim amounts but are not part of the 'amount stated as due' as such. They therefore do not attract retention, which is reasonable as it is already being held on the work carried out and now destroyed. Under clause 22C, the restoration work is deemed to be a variation and as such ranks for retention, even though this is less reasonable.

The piling carried out by the nominated sub-contractor is now complete. A check on the payments made to him has revealed that the contractor is holding a 5% retention, that is the maximum level to which he is himself subject. So far as the contractor is concerned this should now be reduced to 3% as the contract maximum, even though this will mean the contractor paying out more than he receives. (Be it noted in passing that in the closing stages of a contract this position can be reversed.) While no deduction has been made from the valuation, the surveyor has notified the architect, who in turn has certified the lapse to the employer as he is obliged to do. Since it appears to be a purely technical slip in this case arising from the use of a non-standard condition, no direct payment is being made and the contractor has undertaken to remedy the position.

MONTH 11
The results of the explosion have now all been tidied up and work has, in the

main, continued as planned. There was a delay of some six weeks in the arrival on site of the ventilation sub-contractor, due to a late nomination by the architect. As a result, a month extension of time has been granted on the office block only, which is where the ventilation work was first needed. Payments are correspondingly behind. A claim for additional payment under clause 26.1 has been agreed covering labour standing and some time for plant and supervision also affected. There has not been any extra agreed in relation to adjusted preliminaries, since no overall prolongation of the contract period has resulted. Even so, such an extra is conceivable.

On more homely matters, there has been a somewhat tardy inclusion over the last few months of the value of variations carried out to date and also of fluctuations incurred, both resulting in a net addition. Of these, variations carry retention but fluctuations do not (although they would have done if clause 40 had been in operation). Some fluctuation increases were notified late and the architect has refused to admit them, as he would have tried to take action to avoid the increase. The contractor has given notice that he is seeking arbitration over this item after practical completion.

The valuation is now taking on a somewhat more complex appearance, and in full would by now be quite a lengthy document even with any incorporation of previous totals that could safely be undertaken without risk of error. It also allows for a reduction of retention; this is explained immediately after the valuation analysis.

		£	£
Preliminaries:	Early items	92,000	
	Regular items (10,400 × 10)	104,000	196,000
Bills 3 to 6:	As contract	1,824,000	
	Net variations	48,000	1,872,000
Repair work, clause 22C			32,000
			2,100,000
	Water etc. 2/3%, say		14,000
		Total work	2,114,000
	Materials on site		205,000
			2,319,000
Less retention	Standard 5%, say	116,000	
	Less reduction on piling	1,500	114,500
			2,204,500
Gross payments:	Local authority's fees	2,400	
	Repair work, clause 22A	44,000	
	Net fluctuations (including nominated sub-contractors)	88,000	
	Loss and expense: clause 26.1	14,000	148,400
			2,352,900
	Less previous payments		2,085,200
		Valuation 11	£267,700

The piling sub-contractor approached the quantity surveyor through the contractor asking that he might be paid off completely, as his work was obviously satisfactory and yet the latest release date for retention under clause 35.17 was still some nine months away. The quantity surveyor could not act on his own and so he referred the matter to the architect. As the employer/sub-contractor agreement is in force and has been fully observed, the reduction of retention has been agreed and is thus of the whole amount concerned and not just half. The item is best shown separately as a reduction of the maximum amount held at the present stage, so that its occurrence is clear to all reading the valuation and related certificate. The agreed account for the piling is £50,500 and the retention held at 3% maximum is thus £1,500. In passing it may be noted that, as the original PC sum was for £65,000, the contractor will be subjected to slightly more than the expected share of the retention on the remainder of the work when the maximum is reached. On a large account this differential could be quite substantial. Unlike the similar reduction that occurs at partial possession or sectional completion, the present one is in one stage and there is not an accompanying reduction in the level of liquidated damages.

MONTH 14

This is the first of three months for which it is not necessary to give a full analysis of the valuation, but in which points of interest may be noted. The main problem during the month has been a delay in the issue of drawings for the stores extension. As a result the architect has granted a one month extension on that section of the work. The contractor has agreed that this is the correct measure of the delay, but has contended that the whole of the works, except the office block, should receive this extension, since there is only one completion date for all that work. He has stated that he will use his best endeavours to prevent delay, as clause 25 requires him to do, but that he wishes to protect his position against any later claim for damages. The architect has accepted this argument and amended the extension to cover the whole apart from the office block, which retains its own date for completion.

As a result of this delay, the payment of preliminaries has slipped a further month behind for most of the works and is calculated as follows:

	£
Early items	92,000
Regular items, all sections (10,400 × 12)	124,800
Ditto, office block only	2,400
	£219,200

This is only a temporary setback in the formal calculation and is in any case compensated for by the inclusion of an extra sum for prolongation of preliminaries in the claim agreement that has resulted from the delay. The main heads of this claim are as follows:

		£
Preliminaries:	One month of regular items	10,400
	Less office block	2,400
		8,000
Labour and plant standing on stores extension		13,600
Related effect on process building		2,400
Storage of materials at works and double handling		
resulting, mainly for sub-contractors		12,000
		£36,000

In the circumstances that have led to the storage of materials away from site, it becomes inevitable that a payment on account for them should be made under clause 30.3 which applies in this contract under English law. This is done, after the necessary formalities of that clause have been observed. An alternative would be to agree to an interest charge on capital tied up, if this were acceptable to the firms in relation to their liquidity.

This valuation gives a gross value of those items subject to retention of well over £2,899,200. This is the sum on which 5% retention reaches £144,960 which is the contract limit of 3% on the contract sum. Since a reduction of £1,500 was made in Certificate 11, the maximum has never actually been held. From now onwards further sums paid will not attract retention and in terms of arithmetic it does not matter whether items previously shown as gross payments are added in before the retention deduction or after it. For consistency, however, the same order is desirable.

MONTH 17
Early in this month, the employer took possession of the office block by arrangement with the contractor; this was thus within the extended date for completion at the end of this month by a matter of weeks. As this falls within the arrangements envisaged initially, it is sectional completion under the contract supplement. There is thus practical completion, and additionally the architect has issued his certificate of the approximate total value of the part taken over in accordance with clause 18, so that a reduction to half retention can be made.

Here the parties are caught in a web of their own devising. The limit of retention is set at 3 per cent of the contract sum, so that the actual percentage held on the value of work is constantly shifting when once the limit has been passed, while no rules have been spelt out for calculating its reduction. The uneven effect of releasing the comparatively small amount of retention on the piling has already been noted. For the present larger unit, it is agreed to make the reduction proportionate simply to the amount of the contract sum, that is without taking account of variations and other adjustments. Only the totals of the bills for the physical entities are needed, as the other amounts are assumed to be proportionate. The calculations are:

	£
Bill 3: Office Block	880,000
Bill 6: External Works (part taken over only)	40,000
	£920,000

This is approximately 22% of the totals of Bills 3 to 6, i.e. of £4,240,000.

22% of the total retention of £144,960 is approximately £31,900 and half of this is due, i.e. £15,950.

Normally liquidated damages would be adjusted by another calculation in accordance with clause 18, but a fixed amount has been provided in the contract appendix. The problem caused by the formula in the clause will occur shortly, however.

MONTH 19

Two causes of delay come to a head in this month in the course of negotiations. One is a straight delay of two weeks on the part of the ventilation sub-contractor in the stores extension. The contractor has taken all reasonable steps to progress this matter and so there is no real argument over this. No question of claim for payment under the contract arises although the contractor has pursued a charge against the sub-contractor for disruption of work.

The second matter coming to a head this month is a delay caused by the introduction into the process building of several direct contractors to the employer. It had not been expected that they would have appeared until one month from completion, as was stated in the contract bills in accordance with clause 29.1. They have been brought on to the job early in the hope of making up some of the time lost earlier. The contractor agreed under clause 29.2 to them coming on site, but pointed out that they would create congestion for his own workmen. He therefore warned of his desire to be reimbursed for the expense that he would encounter and also to receive an extension of time. This was a prudent course of action, duly confirmed in writing, since the direct contractors were likely to cause a delay whether they worked to their own programme efficiently or not.

On this latter score an extension of time of six weeks is requested. The architect points out that a one month extension applied to the process building in Month 14, when that part of the works had not been delayed. The contractor replies that he has rephased his work to take account of this extension and so that he can keep his labour force working economically between the various buildings. After some discussion an extension of one month is agreed, this including the stores extension. This building is progressing well and thus has benefited by a two weeks windfall. Appendix 1 at the end of this chapter shows the way in which effort has been allocated.

The claim as agreed covers several elements of expense which can be analysed as follows:

	£
One month's prolongation of preliminaries	10,400
Labour standing and working at low productivity	24,000
Repair work due to operations of direct contractors	8,000
	£42,400

MONTH 21

At the end of this month work on the stores extension is going very well and is likely to be completed one month ahead of its date for completion as extended. The process building is slipping behind, even though the effects of the plant installation work have now been overcome. There has been a change of foreman and this is apparently the trouble.

Eighteen months of regular preliminaries are now due, since there has been three months total extension of time and any extra preliminaries have been allowed in the claim settlements. In addition, some of the items allocated to the closing stage should now be coming forward for inclusion. As the works are both ahead and behind in their various parts, a fairly low sum is allowed for these items.

		£	£
Preliminaries	Early items	92,000	
	Regular items (£10,400 × 18)	187,200	
	Office block completion	2,400	
	Further completion	3,200	284,800
Bills 3 to 6	As contract	4,056,000	
	Net variations	144,000	4,200,000
Repair work under clause 22C			32,000
			4,516,800
	Water etc. 2/3%, say		30,100
		Total work	4,546,900
	Materials on site		31,000
Less retention	Maximum	144,960	4,577,900
	Reduction on piling etc. 7,000		
	Reduction on office block 15,950	22,950	122,010
			4,455,890
Gross payments	Local authority's fees	2,400	
	Repair work under clause 22A	44,000	
	Net fluctuations	281,000	
	Loss and expense: Month 11	14,000	
	Month 14	36,000	
	Month 21	42,400	419,800
			4,875,690
Less previous payments			4,669,800
		Valuation 21	205,890

Only a few comments are needed. The 'water etc' allowance is related to the whole of what precedes it in total; since this is a convenient approximation, it will be necessary in these closing stages to watch that the amount is not accidentally exceeded. A revision of it will be a different matter, of course. Materials on site are dropping in value; a check has been made that the amounts included do not include items that are surplus to the needs of the project. It is to be hoped that minor items like these will be absorbed by the retention, but there is no point in being out for the sake of it.

Several reductions of retention for nominated sub-contractors on either the stores extension or the process building are now shown, in addition to that for the piling. No detail is needed for present purposes.

MONTH 24

The project is on the verge of practical completion, but the architect has certified at the end of Month 23 that completion has not occurred at the then fixed completion date. The employer has been advised that he may deduct liquidated damages from the sums falling due to the contractor, but they cannot be deducted by the architect from the amounts otherwise due. For this reason the quantity surveyor must not deduct them in his valuations either. The level of damages is shown in the supporting calculations following the valuation itself.

The stores extension was, however, taken over by the employer under the partial possession (rather than sectional completion) provisions at the end of Month 22 and this has led to a further reduction of retention and a reduction in the level of liquidated damages now chargeable. The certificate of making good of defects in respect of the office block is being issued at the same time as Interim Certificate 24. The second half of the reduction of retention must therefore be made.

There has been a difference of opinion with the architect over liquidated damages on the part of both parties to the contract. The contractor has contended that as he handed over the stores extension early he should have this offset against his lateness on the remainder of the works. It has been pointed out that the benefit he receives for this is the early reduction of retention and the ending of several of his liabilities early.

The employer has complained that the process building, which is now late, is worth more to him, in proportion to its capital cost, than is the stores extension. His suggestion that the liquidated damages should be split in some other proportion than that resulting from the formula in clause 18.1.5 has, however, been declined.

On the office block a small amount of damage has been caused by a storm. This has been met out of the insurance taken out by the employer at the time of sectional completion, when the contractor ceased to be responsible. The contractor has accepted the work as a variation, as the amount is quite small, although he could have declined it if he had wished.

The valuation can be given with some abbreviation of detail:

		£	£
Preliminaries, entire except for some removal of plant and other minor work			310,000
Bills 3 to 6	As contract	4,240,000	
	Net variations	140,000	4,380,000
Repair work under clause 22C			32,000
			4,722,000
	Water etc. allowed complete		32,000
	Carried forward overpage		4,754,000

				£
Total work (brought forward from previous page)				4,754,000
Materials on site				2,000
				4,756,000
Less retention	Maximum		144,960	
	Reductions: piling etc.	7,000		
	office block	15,950		
	stores extension	19,750	42,700	102,260
				4,653,740
Gross payments	As month 21		419,800	
	Further fluctuations		17,500	437,300
				5,091,040
Less previous payments				5,021,020
		Valuation 24		70,020

The value of variations has been reduced slightly now that rates and measurements are closer to finality. Any figures in an interim payment are provisional and may be revised in this way if necessary, by virtue of clause 30.10. 'Water etc.' has been given at the contract amount in full; if it is going to be adjusted it will be upwards. This will be done in relation to the competitive items in the contract bills.

A closer estimate than that available in Month 17 is now to hand for the office block but the figure already included in the architect's certificate as being the value for the purpose of sectional completion must stand contractually. To arrive at the reduction for the stores extension a corresponding exercise was performed as follows:

	£
Bill 4: Stores extension	1,200,000
Bill 6: External Works (part taken over only)	60,000
	£1,260,000

This is approximately 30 per cent of the totals of Bills 3 to 6 as used before and gives a corresponding amount of retention of £43,500.

There is a further point to watch at this stage in calculating the reduction:

	£
Total retention on Stores Extension	43,500
Less reduction on sub-contractors already made in full for this building	4,000
	£39,500

Half of this reduced total is due at this stage, i.e. £19,750.

A distinct calculation is needed to arrive at the reduction of liquidated damages due to partial possession. In this contract a fixed division of damages applies between the office block and the other buildings, and the share for the office block has already been released under the sectional completion supplement. There is now a reduction of part of the other fixed amount under clause 18 to be made. To permit this, the first step is to allocate the contract sum between the two parts with the fixed shares of the damages. This is done as follows:

		£
Contract sum		4,832,000
Less Office Block and External Works, as for retention	920,000	
preliminaries, water, etc. 8.3%	76,000	
contingencies, 5.66%	52,000	1,048,000
		£3,784,000

This gives the total value of work that is subject to the £8,000 share of damages.

In these calculations the external works have been allocated according to their date of handing over only, and not therefore according to any formula based on the value of the buildings and the independent or shared use of elements of the external works.

To segregate the stores extension from the above figures, it is necessary to observe clause 18 carefully. This requires that the reduction of damages for any part taken over shall be in proportion to the 'value' of the part to the contract sum. This value will therefore include the net adjustment for variations, while the contract sum will include the contingency sum. These differences may or may not tend to balance out overall, but they do create anomalies for the calculation of the various parts. The last part remaining as *not* taken over is valued as the residual part of the contract sum after subtracting the earlier parts with their variations. The last part thus includes the whole amount of the contingencies (rather after the manner of a game of 'pass the parcel') as well as its own share of variations. Whether the 'value' of a part taken over should include amounts for fluctuations, loss and expense and repair work under insurance (all of which have occurred in this case) is arguable. They do not affect the 'value' at the contract price level and result in no physical change. Also, the same estimate of value is to be used as the basis of the reduction of both retention and damages, so that it may be assumed that they are to be ignored, at least so far as they are net. Even the repair amounts can be ignored here because of the limit on retention. This appears to be the best interpretation of an awkward clause, even though it might be agreed that some part of contingencies has been included to provide for just such costs.

The starting figure is that just used for the reduction of retention. It must be adjusted to include elements not needed for the previous simple proportioning. In this case the result is not to be used for a slowing down of cash flow, but for a possible permanent withholding of money; but the developed basis is needed to produce the result at all.

	£
Stores Extension and External Works	1,260,000
Preliminaries, water, etc. 8.3%	105,000
	1,365,000
Variations net addition (including any adjustment of preliminaries)	37,000
Total value of work not now subject to part of £8,000 damages	£1,402,000

Allocation of damages to stores extension and not now applying

$$8,000 \times \frac{1,402,000}{3,784,000} \text{ say, £2,964 reduction per week}$$

The balance of £5,036 per week is therefore the sum attributable to the process building and has been advised to the employer

The figures that would have applied on a straight proportionate basis are:

Stores Extension	£3,020
Process Building	£4,980

The process building is therefore carrying 'excess' damages of £56 per week, and this would have been higher had the damages for the office block been fixed as high as those for the rest of the contract in relation to value and had no fixed allocation applied. This would have been mainly due to the passing forward of the parcel of the preliminaries.

Looking forward

At this stage the clock stops, but a forecast of the future may be made. Practical completion is due in about two weeks and marks the end of the contractor's liability to perform fresh work. A small shed is needed by the employer, however, and the contractor has agreed to erect it provided it can be completed during the defects liability period. He will be paid at contract rates plus 35 per cent and a special item for supervision.

Following the certificate of practical completion, there will be an interim certificate including the rest of the first half of the retention reduction. Other reductions will follow as they arise and there will be one or more further interim certificates to pay off any further values accruing at not less than monthly intervals.

Also as a result of practical completion, liquidated damages will cease to accrue, although those already due may be deducted at any time until the final certificate, which should take account of any not then deducted. There may also be the need to adjust the numbers of weeks of damages payable when the architect ultimately reviews the extensions of time granted during progress. There may therefore be the possibility of a refund of damages to the contractor. All of this however lies outside the final account itself. The arbitration over the fluctuations dispute may proceed, some preparatory work having been undertaken by both parties.

A glance at how the final account is likely to look may be permitted, leaving out only the amounts for the shed mentioned above. It will not be in this form, but the figures below work progressively from the original intention through the several layers of extra expenditure incurred by the employer:

	£	£
Contract sum, less contingencies		4,592,000
Amounts to set against contingencies		
Variations	140,000	
Local authority's fees	2,400	
Loss and expense	92,400	234,800
Total A		4,826,800
Fluctuations		298,500
Total B		5,125,300
Insurance		76,000
Total C		£5,201,300

Total A represents what should be set against the contract sum in viewing how good a steward the architect has been. He has done well, so long as some of the detail is not probed! Total B was envisaged in principle by the employer and the level of it is largely outside the control of the parties and their advisers. Total C was not expected at all, but the sums causing it to occur have been met by others than the parties to the contract. Beyond all these figures are the delay losses to be met by the employer or recovered in damages from the contractor.

How the contractor has fared throughout has not been the centre of concern in this study. He is left looking at a moderately favourable forecast of the final settlement. He is wondering whether he can successfully found a claim for an overall disturbance of his regular progress that has not already been absorbed within the amounts for separate disturbance. He is not too sure, since he has agreed the other sums in fairly definite terms, and he has not yet given notice of any further effect. He is doubting whether such a notice will be within the reasonable time that the contract requires. His doubt is probably justified, but there may just be a hope. At any rate the final account rates are not yet all cleared and while there is life there is hope. On this optimistic note it is as well to end.

Appendix 1
Anticipated rate of expenditure

The table below gives the rate of expenditure anticipated had the project run entirely to its original programme. It did so for the first six months and was then delayed several times and the expenditure shown in Appendix 2 occurred. The figures are taken solely from the bills concerned, with no addition for preliminaries or the like and no final account adjustments. All figures represent thousands of pounds, except the month number.

Month	Bill 3 Offices	Bill 4 Stores	Bill 5 Process	Bill 6 Externals	Totals
1	24	16	40	24	104
2	24	24	56	24	128
3	40	24	64	16	144
4	48	40	64	24	176
5	48	40	80	—	168
6	56	64	80	—	200
7	56	64	96	—	216
8	80	64	96	—	240
9	96	56	112	8	272
10	80	56	112	8	256
11	104	64	128	—	296
12	80	80	136	—	296
13	64	80	136	—	280
14	48	88	136	16	288
15	32	88	128	16	264
16	—	96	112	16	224
17	—	96	96	24	216
18	—	72	96	16	184
19	—	56	104	24	184
20	—	32	48	24	104
	880	1,200	1,920	240	4,240

The actual month-by-month expenditure over the 24 months taken to complete the contract is shown in Appendix 2 which follows.

Appendix 2
Actual rate of expenditure

This table gives the actual expenditure that was incurred each month after Month 6, again based solely on bill figures. Where there was an extension of time granted, this is indicated by an asterisk. The figures are again representative of thousands of pounds.

Month	Bill 3 Offices	Bill 4 Stores	Bill 5 Process	Bill 6 Externals	Totals
1–6	240	208	384	88	920
7	8	16	16	—	40
8	48*	48*	80*	—	176
9	80	64	96	—	240
10	—*	56	112	8	176
11	96	56	112	8	272
12	80	64	128	—	272
13	104	40	136	—	280
14	80	40*	136*	—	256
15	64	80	136	16	296
16	56	96	112	16	280
17	24	104	96	16	240
18	—	112	80	24	216
19	—	48*	24*	16	88
20	—	64	72	8	144
21	—	72	72	16	160
22	—	32	56	16	104
23	—	—	40	8	48
24	—	—	32	—	32
	880	1,200	1,920	240	4,240

It will be seen that the effects of delays on the expenditure rate of the parties varies according to the timing of delays within the overall period and also between the buildings affected. Changes in rate can affect each party's wider financing pattern.

CHAPTER 25

Case B: A contract interrupted

By comparison with the previous case, and with apologies to the more scholastic, this case study might be sub-titled 'per ardua ad disastra'. It is interrupted by that most dreaded but alas all too common event in the construction industry—insolvency of the contractor. Signs that all is not well appear as early as Month 4, although the contractor does not actually go into liquidation until Month 8. The sad story makes its way through several stages:

(a)   Before insolvency the main actions are as laid down by the contract established between the parties. In the circumstances that develop the virtues of careful conduct on the part of the employer's advisers become plain.

(b)   When the determination comes about a number of actions fall to be taken. Some are for directly physical reasons affecting the site, while others are financial and have their own flavour of risk so far as their outcome goes.

(c)   The contractors for completing the work are fairly placid, although they have their day to day problems in tidying up what was left behind.

(d)   Settlement of the account with the original contractor leads to measures of financial artistry.

(e)   As a sideline, the position of the largest nominated sub-contractor at the various stages is also examined.

As was done with the previous case, a number of relevant details are set out first and again there is an analysis of anticipated expenditure, given in the appendix to this chapter. Many of the figures that follow are set out to show the points under discussion at their plainest and are not presented as they might well be in their full context.

In particular, discounts on nominated accounts have been ignored for the sake of easy working; the contractor would retain these if he paid promptly while the employer would pass them on to the firm when making a direct payment in the case of the contractor defaulting. In the case that occurs here, if the employer pays an amount for retention direct to a firm after the contractor has had his employment determined the employer would retain the discount on the retention, as he was the person paying on time; a payment through the liquidator would change this.

It has been assumed that nominated work has been let at the sums included in the bills, that it has not been subject to variations and (except in the case of the precast frame) that it has not been subject to fluctuations.

Interim payments made have also been assumed to have been accurate payments for most final settlement purposes, since this makes the figures easier to trace through the various stages.

Some aspects of the contract

The only building in this contract was a private youth centre. It would seem that the management committee showed undue zeal in securing a low initial tender at the expense of what was to follow. A selective analysis of the contract bills is as follows:

		£	£
Preliminaries	Early items	14,000	
	Regular items	24,000	
	Late items	3,000	41,000
PC sums	Nominated supplier decking		
	slabs	9,000	
	Nominated sub-contractors		
	precast frame and panels	84,000	
	roofing	12,000	
	heating	54,000	
	others	36,000	195,000
Builder's work	Building	156,000	
	External works	20,000	176,000
Contingency sum			20,000
	Contract sum and total of bills		£432,000

Full fluctuations as clause 39 were allowed for in the conditions and it was stated in the preliminaries that payments to the precast frame sub-contractor might be made for materials off-site, in accordance with clause 30.3, provided that a particular procedure was followed in addition to that in the clause.

The points of interest in the appendix to the conditions are:

(a) Date for possession   Date Y
(b) Date for completion   Date Y plus 15 months
(c) Liquidated and ascertained damages   £300 per week
(d) Retention percentage   5%

The high proportion of early items in the preliminaries may be noted as having an unfortunate effect for the employer.

No further analysis is needed in this instance and so the story may be sampled at its most significant moments.

Position at end of selected months
MONTH 5
Work on site began steadily enough and held reasonably to programme for the first two months. By the end of this present month it was running about a month and a half behind and it was beginning to look already as though the contractor would not be of the calibre to regain this lag. The precast frame sub-contractor was being held up but had not yet asked for any payment for materials fabricated but not delivered. He had, however, complained that £1,000 of the sum involved in the certificate at the end of Month 4 had not been paid. The contractor attributed this to some contra-charges not established by the defined procedure of the standard sub-contract in use. The employer was advised when the present certificate was issued that this had occurred and told that he had an option to make a deduction and pay direct. Direct payment was not obligatory, as the employer/sub-contractor agreement was not in use, despite the design element in the sub-contract. Acting on the architect's advice, the employer did not make the deduction but waited to see how things might develop. The contractor was told that he should pay the sum and either render a separate account for the contra-charges or deduct them when finally agreed.

In the valuation that follows, the level of preliminaries has been held at that included in Valuation 4, to allow for the delay.

		£	£
Preliminaries	Early items	14,000	
	Regular items (1,600 × 4)	6,400	20,400
PC work	Precast frame		21,000
Builder's work	Building	36,000	
	External works	6,000	42,000
	Total work		83,400
Materials on site	General	6,000	
	Decking slabs	3,000	
	Precast frame	12,000	21,000
			104,400
	Less retention 5%, say		
	Contractor	3,550	
	Precast frame	1,650	5,200
			99,200
Less previous payments			72,100
	Valuation 5		£27,100

The employer has already had cause to make deductions from earlier amounts stated as due under interim certificates. One of these was due to the failure of the contractor to take out proper insurance as required by clause 22A of the conditions and to the employer doing this in his place. The other was due to the contractor refusing to remove insitu concrete work when so instructed by the employer before the precast frame sub-contractor could proceed. Another firm was engaged by the employer to do this work, after due notice had been given to the contractor. The amount of previous payments to the contractor shown in the present valuation has not been reduced to allow for these payments, since they have been made by the employer although not through the contractor. If they were deducted now, they would increase the present payment and thus result in a total overpayment. Equally, the sums do not occur in the final reckoning later in the chapter. The architect has, however, cleared the position with the employer to avoid any misunderstanding of the figures at a later stage.

MONTH 6

In the past month progress has been rather better, although none of the lost programme time has been regained. In the valuation the allowance for preliminaries has been retarded by one month and is still below what it should be, if work were on time.

Payments to the precast frame sub-contractor are now behind to the extent of £12,000 net and the sub-contractor has suspended deliveries, while work on site has been slowed down. He has asked that all future amounts should be paid to him direct, but the contractor has refused to agree to this even if the discounts are retained for him by the employer. It would appear that the contractor regards the monies as a useful way of funding his operations, provided he can make all his payments somewhat behind time. It has been agreed that materials off-site will be included in next month's certificate. This gives the sub-contractor a degree of protection in that he will not be risking having materials on site and not paid for at the time of any determination. It does not protect him against the sum included in a certificate not reaching him, however, once the certificate has been honoured. At this point, a sub-contractor is in a better position with the Scottish direct purchase arrangement.

In the meanwhile, the employer has agreed to pay direct the default of £12,000 and he will therefore deduct it in making the payment arising out of Valuation 6 below and the resulting certificate. He had wanted to pay the further sums in the present certificate direct also, out of concern for their fate, but has been advised that the architect cannot certify this and that he cannot act without a certificate.

The decking supplier has not yet been paid, it appears, but here the contractor is still within the days of grace allowed, since these materials were delivered only during the preceding month. It may be noted that the contractor's own materials are low in this month. Little fresh stock has arrived.

When the architect issues his interim certificate to the employer, as given below, he at the same time sends him another certificate notifying the non-payment of the precast frame sub-contractor.

		£	£
Preliminaries	Early items	14,000	
	Regular items (£1,600 × 5)	8,000	22,000
PC work	Decking slabs	3,000	
	Precast frame	42,000	45,000
Builder's work	Building	39,000	
	External works	6,000	45,000
	Total work		112,000
Materials on site	General	3,000	
	Decking slabs	4,400	
	Precast frame	6,000	
	Heating	3,500	16,900
			128,900
	Less retention 5%, say		
	Contractor	3,900	
	Precast frame	2,400	
	Heating	200	6,500
			122,400
Less previous payments			99,200
	Valuation 6		£23,200

If this valuation were prepared without reference to that preceding it, its significance might not be obvious. Much of the progress that has been made is on the part of nominated firms, before the real state of the contractor began to penetrate in their awareness. In fact, out of the net sum payable about £21,750 is on their behalf. The contractor has done some work but this has been cancelled out in overall effect by the drop in materials on site and a tightening of the quantity surveyor's calculation in general. Only the allowance, in a rather generous moment, of a full month's preliminaries has advanced things at all. These add up to a string of danger signs when taken alongside the failure to pay the accounts due.

MONTH 7
Progress has been poor during the last month. Again it is largely to be explained by the value of nominated work and materials, especially off-site value, in the current valuation. The contractor's own work appears to show an increase but this is illusory in the sense that a firm to whom work has been sub-let has been active on site.

The signs of the contractor being in real trouble are unmistakable. There are rumours of unofficial meetings of creditors and that official actions are under way. Progress on site is now fully two months behind and the stock of materials is even lower. The decking slabs supplier has complained that he has not been paid, but has been advised that the employer is powerless to protect him. The conditions provide for payments to be made but give no sanctions to employ if they are not. The supplier has now suspended both credit and deliveries. The precast frame sub-contractor has not been paid any more and his position is set out below, following the valuation detail. The heating sub-contractor has received nothing at all. Deliveries of roofing materials are due, but are 'unaccountably' delayed.

All that the quantity surveyor can do is to warn the architect and through him the employer of any trends and rumours and prepare the valuation as usual, with especial care over all doubtful points. It is as well to avoid any suggestion of panic that might precipitate collapse. He has decided to hold the preliminaries at the same sum as for the previous month, as progress is now correspondingly behind, despite the risk of killing off the contractor. At this juncture, the employer's best interests are served by caution in this way. (So too are the surveyor's!)

		£	£
Preliminaries	As month 6		22,000
PC work	Decking slabs	3,000	
	Precast frame	48,000	
	Heating	2,000	53,000
Builder's work	Building	42,000	
	External works	6,000	48,000
	Total work		123,000
Materials	General	1,600	
	Decking slabs	4,400	
	Precast frame		
	on-site	6,000	
	off-site	6,000	
	Heating	2,000	20,000
			143,000
	Less retention 5%, say		
	Contractor	3,900	
	Precast frame	3,000	
	Heating	200	7,100
			135,900
	Less previous payments		122,400
	Valuation 7		£13,500

This valuation does not deduct the amount of £12,000 paid to the precast frame sub-contractor by the employer last month, any more than have deductions been shown for the earlier direct payments. In fact, there is no strict need for the surveyor even to know that such a payment has been made, other than to enable him to check the correctness of the overall payment made by the contractor to the firm. Even this is contractually the architect's duty.

When the certificate is issued, the employer has two problems to resolve. One is whether to pay quickly and thus help to keep the contractor solvent, or whether to delay payment as long as possible and perhaps not have to make it at all if the contractor becomes insolvent before the period for honouring the certificate runs out. This latter course may however precipitate the very insolvency. The problem is not contractual, but rather moral and a matter of business expediency. It is overlaid in this instance by the second problem of payments to the sub-contractors, or the lack of them. The precast frame sub-contractor may be considered further here.

His position is unenviable and could always lead to him going the way that the contractor is heading. He has been paid £12,000 direct by the employer, but the contractor has now defaulted over a further £14,250 and there is more in the current certificate. (It would have been less risky for him, as mentioned, had he not asked for the off-site portion to be included, but he was hoping to help his own cash flow.) He has therefore prudently stopped effective work by this stage. The full story of his account is given at the end of this chapter. The employer can do nothing about the last amount at present and there may be no future for the contractor. The £14,250 now due is in excess of the amount due at present to the contractor and so to pay the whole amount direct will be to incur an over-expenditure on the works to date. Furthermore the heating sub-contractor is short by £3,300, even if he has joined the queue more recently.

To pay the full amount of the current certificate between the two sub-contractors is the most that can be done and this has the risk of impoverishing the contractor completely. Legally, the employer need pay the sub-contractors nothing in the absence of collateral agreements, but he wishes to protect them and decides, with advice, to pay a total of £9,000 and give the meagre balance to the contractor. The amounts are based on clause 35.13 by being proportionate:

	100% Default	*77% Available*	*51% Paid*
Precast frame	14,250	11,000	7,300
Heating	3,300	2,500	1,700
	£17,550	£13,500	£9,000

The architect has had to follow a strict line in these deliberations. He must issue Certificate 7 and include the full amount now due to the sub-contractor and thus 'direct' the contractor to pay in accordance with clause 35.13. He must also certify to the employer the default over last month's payment. He may advise the employer informally over whether to delay honouring the certificate for as long as possible or whether to pay as quickly as possible to keep the contractor going. He dare not suggest to the employer that he delay honouring for any period in excess of that laid down or the contractor may be able to determine against the employer on quite favourable terms to himself.

MONTH 8

The contractor has gone into liquidation just two days before Certificate 7 was due to be honoured in the reduced sum of £4,500. The direct payment in total to the sub-contractors of £9,000 has however been made, even though it should only have happened at the same time. Upon hearing that the formal winding-up proceedings were under way, the employer settled for the maximum time before paying the contractor and has just made it. The payment now lapses and will help to cushion the blow of a break on site. The liquidator has agreed to accept an automatic determination under clause 27.3 and not press for a reinstatement. Just after the determination, the roofing sub-contractor rather unexpectedly delivered a consignment of materials to site.

A review of the financial position has been asked for to see what is in effect in hand. A balance between payments and physical assets, using the prices of the contract, is as follows:

	£	£
Gross amount of Certificate 7		143,000
Undervaluation of work		1,000
Fluctuations not claimed		2,000
Roofing materials not included		2,000
Work etc. on site		148,000
Less		
Net amount of Certificates 1 to 6	122,400	
Direct payment at time of Certificate 7	9,000	
Sub-contract materials off-site	6,000	
Sub-contract retention from Certificates 1 to 6	2,600	140,000
Notional amount in hand		£8,000

This amount is all that is available from the contract monies to help towards finishing the works and it is not going to be any too much as there is still a large amount of work to be performed. The retention shown on the sub-contractors has to be met by virtue of trust fund status. That from Certificate 7 never came into existence as part of the fund.

A reconciliation of the reasons for this notional £8,000 is:

	£
Main retention from Certificates 1 to 7	3,900
Sub-contract retention from Certificate 7	600
Payment not made, net	4,500
Undervaluation, fluctuations and roofing delivered	5,000
	14,000
Less materials off-site	6,000
	£8,000

The contractor appears, from all accounts that have come to hand, to have failed to make the following main payments:

	£	£
Precast frame as detailed in the closing part of this chapter	18,350 6,000	12,350
Decking slabs		7,400
Heating		1,800
Merchants		3,000
Sub-let work		1,600
Hire of plant		1,400
		£27,550

In addition, the roofing has been delivered since and the firm is outstanding on its value, although the contractor's liquidator is perhaps not liable for it. The employer is under no obligation to pay any of these amounts and may find himself paying twice for them if he does by virtue of clause 27.4.2, (see *BCPG*, Figure 12). Even if this does not happen, anything paid will reduce the monies available for completion and widen any gap left at the end. Such a gap will become a sum to be sought from the liquidator at whatever rate of settlement is available. Payment should therefore be avoided where possible. In due course the following partial payments are made as agreed with the liquidator as a protection against making actual or alleged double payments and all net of retention:

	£
Precast frame; since this item is a monopoly as a system and a fresh sub-contract must be arranged	4,400
Decking slabs, again to ensure continuity, since the original price was very favourable	3,600
Roofing, since the delivery was late	1,600
	£9,600

These payments represent a curious blend of expediency and kindness. Whatever the reasons, they are as effective in eroding the monies in hand. The rest of the balances were put to the liquidator and fall out of the narrative. The hired plant was removed to avoid an implied hiring arising.

Apart from the financial actions that are taken or set in motion at this early state of the determination, several other steps are necessary. The employer has already had to pay the insurances under clause 22A in the contractor's default. He must now take up with the insurer the question of any amendments necessary to cover the period during which the works are to stand idle and also the revision to transform the policy into a clause 22C cover for the completion contracts following. A system of security has to be set up to protect the site and a number of other items cleared from day to day, including attendance during the removal of hired plant. An extra cost of £1,400 is incurred. There was also a small matter of arbitration raised during the early stages; it may proceed now if either party wishes to pursue it.

MONTH 9

With the initial emergency actions taken, it becomes possible to get activity going again on the site. It has been found that the site and the works are both in an untidy state, with materials littered about and many items of work half done and out of sequence. As a result a prime cost contract is let to tidy things up to a clear stage. This gives a final account of £3,000 of which half is for new work and the other half for tidying up activities that would have been carried out by the contractor without further charge had he continued.

Preparation of fresh bills of quantities has started. These can be relatively tidy as the prime cost contract is taking up the worst of the confusion, but they include a fairly large allowance for unforeseen snags left by the original contractor. They allow for clause 22C insurance and also make special arrangements over retention levels on those nominated sub-contracts that are to be brought forward from the original contract, in view of what has been held on them already.

The completion contract

The contract with the new contractor ran through without serious trouble
and may be shown for present purposes in just a few elements. First there is
a synthesis of the contract sum, not as it was but shown as a reconstruction
by adjusting the contract sum for the determined contract:

	£	£
Original contract sum		432,000
Less contingency sum	20,000	
gross payments made	143,000	
undervaluation	1,000	
roofing material (full amount)	2,000	
prime cost contract (half only)	1,500	167,500
		264,500
Variations already ordered on original contract		6,000
Materials off-site		6,000
Higher pricing		
fluctuations equivalent		8,000
smaller contract and working in partly		
completed building		5,000
Setting up the site		7,300
Contingencies		14,000
Contract sum		£310,800

Of the above sums, those for the smaller contract and for setting up
represent expenses that the employer would not have borne under an
undisturbed programme. For simplicity the sums for prime cost work and
variations have been shown at actual costs rather than the slightly lower
figures that would have been incurred in the original contract. The
inaccuracies cancel one another out.

Secondly there is an analysis of the final account at its own price levels,
selecting only a few figures for later use:

	£	£
Contract sum		310,800
Less contingency sum		14,000
		296,800
General variations ordered on this		
contract, net additions	8,500	
Corrections to original contractor's		
work and related variations	4,400	12,900
		309,700
Fluctuations		10,000
Total account		£319,700

The above account thus includes some items that do not show as variations within it but are variations on the original scheme. It also includes some variations that have arisen as a result of replanning part of the external works as the most expedient way of overcoming some of the bad work left as a legacy by the original contractor. These are not split in the account into one element of variation and one of corrective work, since they have not arisen as distinguishable entities. It is a matter of philosophy as to how they should be treated in the later reconciliation, as they would not have happened apart from the determination. But they cannot be ignored. Fluctuations are much lower than they would have been, since they have been largely absorbed in the contract price.

The administration of the completion contract and other work has involved the following additional fees, all charged on a time basis:

	£
Preparing amended bills and agreeing the final accounts etc. required (net extra time only)	4,000
Corresponding architect's services	2,000
	£6,000

The employer's loss and damage

The employer has suffered a number of expenses which are the results of the late completion of the works and not part of the cost of completing it. These would have been covered by the liquidated damages in the original contract had the contractor's employment not been determined. As it has, the actual loss has to be calculated instead and put forward to the liquidator. This is done by the employer and his accountant and amounts to £400 per week over 28 weeks, giving a total of £11,200. The financial basis adopted in this case is to evaluate the cost of remaining in the old premises. One building is owned and captial is tied up, the other is rented and the rent is a direct loss. There are also higher running costs and the travelling time and expenses of staff between buildings. The time calculation is as follows:

To determination of original employment			32 weeks
From determination of original employment to commencement of completion contract			14 weeks
To completion of 43 weeks of work under original programme, which was actually running 10 weeks late		51 weeks	
Less extension time granted	2 weeks		
liquidated damages time	2 weeks	4 weeks	47 weeks
			93 weeks
Less original contract period			65 weeks
Overall delay			28 weeks

This calculation allows the original contractor the benefit of two weeks extension of time which it must be assumed that he would have received, unless they were due to some peculiar circumstance in the late period of time. The delay due to the completion contractor's fault is also deducted, as this cannot be attributed to the original contractor and damages have already been collected once.

The hypothetical final account

All the historical elements are now to hand ready for incorporation into the final settlement of the original contract. What is now needed to complete the final picture is the expenditure which the employer might reasonably have expected to incur. This must mean a certain amount of reconstruction of events in a different setting and the re-pricing of much work at different rates from those used for its payment. Some daywork or prime cost work will become measured work, while other items will be excluded altogether as being for work that the original contractor would have had to carry out free of charge. The bare bones of what happens is as follows:

	£	£
Original contract sum		432,000
Less contingency sum		20,000
		412,000
Net variations (these are shown here at the price level of the completion contract for easy comparison with earlier figures and a deduction is then made to show the price level of the original contract)		
Included in completion bills	6,000	
General, and ordered later	8,500	
Extracted from corrective work	1,400	
	15,900	
Deduct for original price level	1,000	14,900
Fluctuations (these are all allowed at the original margins and include those on the frame sub-contract)		
In original contract period	2,000	
In completion bill rates	8,000	
In completion final account	10,000	
	20,000	
Less excess due to prolongation and not covered by loss and damage amount	1,600	18,400
Total account		£445,300

The settlement achieved

The totals of the various sets of figures that have gone before come forward, along with a few subsidiary figures, to enter into the striking of the final balance.

	£	£
Payments made under the original contract and in its wake		
Net amounts of Certificates 1 to 6 and direct payments		122,400
Direct payment, at time of Certificate 7		9,000
Payments to nominated firms after determination for work and material before or at determination		9,600
Security and other work at determination		1,400
Retention released on sub-contractors for work before determination		2,600
Expenses in completing the works		145,000
Price cost contract	3,000	
Completion contract	319,700	
Additional fees	6,000	328,700
Employer's loss and damage due to delay		11,200
Total incurred		484,900
Less hypothetical final account of expense that should have been incurred		445,300
Net indebtedness of the contractor		£39,600

In the light of these figures, the notional amount in hand as a buffer as calculated in Month 8 looks a little wan. It does not exist to be set against the final debt above, since it has been swallowed up in the preceding figures already. It should be added to the debt to show a gross gulf due to the insolvency as follows:

	£
Net indebtedness of the contractor	39,600
Notional amount in hand	8,000
	£47,600

Since the various sums that have led to this total are scattered about in what has gone before they may be brought together here as a final reconciliation:

	£	£
Transition items		
Payment to frame sub-contractor	4,400	
Payment to roofing sub-contractor	1,600	
Payment to decking supplier	3,600	
Security	1,400	
Prime cost (half)	1,500	12,500
Completion contract		
Price level	5,000	
Setting up	7,300	
Corrective work	3,000	
Variation price level	1,000	
Extra fluctuations	1,600	17,900
Other expenses		
Extra fees	6,000	
Loss and damage	11,200	17,200
Gross gulf due to insolvency		£47,600

This is salutary reading to take to heart, as the main narrative ends. The employer is left in possession of a brand new youth centre but is lining up with the rest of the creditors to receive a settlement of 25% of £39,600 which the liquidator can foresee in his more optimistic moments.

As a footnote, it may be pointed out that all these figures have been prepared to satisfy the contract settlement without the need to draw up a precise final account for the value of work carried out up to determination at original contract rates. Even the notional amount in hand does not appear as such.

The accounts of the precast frame sub-contractor

To simplify the presentation of what has gone before the minimum of detail has been given about this major sub-contractor, whose work formed a large part of the physical structure and whose fortunes were so affected by the events that took place. It is useful to retrace his part as an addendum here.

His sub-contract sum was £84,000 and payments were subject to retention (in what follows, discount deductions have been ignored). In Months 2 to 4 amounts were included without any trouble arising, apart from the nebulous contra-charge which came up when a check was made in Month 5 and which is absorbed in the later underpayments. The monthly position up to determination is then as follows:

	Total included in current valuation £	Payments made for previous inclusions		
Month 5	33,000	Already behind by £1,000 for a		
Less 5%	1,650	contra-charge, which should be		
	———	established as a separate account.		
	£31,350	Architect cannot judge the issue.		
			£	£
Month 6	48,000	Total due		31,350
Less5%	2,400	By contractor		19,350
	———			———
	£45,600		Gap	£12,000
Month 7	60,000	Total due		45,600
Less 5%	3,000	By contractor	19,350	
	———	By employer	12,000	31,350
	£57,000			———
			Gap	£14,250
Month 8		Total otherwise due		57,000
The previous certificate		By contractor	19,350	
was not honoured and		By employer	12,000	
the amount shown		By employer	7,300	38,650
lapsed.				———
				18,350
		By employer after		
		determination	4,400	
		By employer in		
		completion contract	6,000	10,400
			Gap	£7,950

The sub-contractor is thus left looking to the liquidator of the contractor's affairs for the balance left above and for a loss and expense amount. He is also losing out to the extent of £2,800 for fluctuations incurred before the determination which he omitted to bring forward for interim payment. This too he must present to the liquidator, as he would have had to have done with any further amounts undercertified. However, he is in a better position than he might be, by virtue of the employer's actions. He now has to make the best of what follows.

A fresh nomination has to be made to the completion contractor, the original sub-contract having been automatically determined along with the main contract. Due to his monopolistic position, the sub-contractor is able to give a fresh quotation for the balance of the work, which is higher than the original balance. This is partly to be expected, since it includes some rises in prices that would have been fluctuations under the previous arrangement and allows for a break in the work and a smaller amount of work under the new sub-contract. It is more than these considerations might have suggested, and undoubtedly allows a sum to offset some of the loss that is being suffered. Negotiation from a position of weakness by the surveyor can do little in the circumstances, other than mutter about goodwill. The amounts finally received by the sub-contractor from all sources are as follows:

	£	£
Paid during original contract		38,650
Paid on determination		4,400
Retention paid by employer		2,400
Recovery from liquidator for		
Payments gap	7,950	
Loss and expense	1,650	
Fluctuations	2,800	
25% of	£12,400	3,100
		48,550
Completion monies		
Work outstanding at original terms	30,000	
Additions made in requoting		
Fluctuations equivalent	800	
Cost of break	800	
Recouping loss	1,400	3,000
Fluctuations incurred		800	33,800
Total received		82,350

The employer has paid all the above amounts other than that met by the liquidator and the various accounts thus include £79,250. This outcome compares with the original expected outcome and the actual excess incurred:

	£
Original quotation	84,000
Fluctuations, net addition	4,400
Total anticipated	88,400
Cost of break in operations	800
Loss and expense	1,650
Total incurred	£90,850

The loss of £8,500 (the difference between total received and total incurred) will be borne entirely by the sub-contractor, although it has helped by an overlap to reduce the loss suffered by the employer. This is 75% of the sum of £2,800 not included in time for early fluctuations. It is in the interests of sub-contractors to keep their accounts pretty short when contractors are in financial trouble.

Appendix
Anticipated rates of expenditure

The table below gives the rate of expenditure anticipated had the project run to finality under the original contract and at the intended rate. It lost momentum as from the second month and stopped early in the eighth, quite an amount short of the figures below. As in the appendices to the previous chapter, the figures do not include preliminaries and all, except of course the month number, represent thousands of pounds.

Month	Precast frame	Decking slabs	Roofing	Heating	Other PC sums	Building	Externals
1	—	—	—	—	—	6	1
2	6	—	—	—	—	12	1
3	12	—	—	—	—	12	2
4	16	2	—	—	—	6	2
5	18	4	—	2	—	6	—
6	14	3	2	2	—	12	—
7	10	—	4	4	—	12	—
8	6	—	4	6	2	8	—
9	2	—	2	8	4	16	—
10	—	—	—	10	4	16	—
11	—	—	—	9	6	14	—
12	—	—	—	7	6	12	2
13	—	—	—	4	6	12	6
14	—	—	—	1	5	6	4
15	—	—	—	1	3	6	2
	84	9	12	54	36	156	20

Had the slow-down and stop occurred about four months later, the frame sub-contractor and decking supplier might have extricated themselves happily. Other sub-contracts would have been placed and operating, just as the roofing and heating sub-contractors in fact were beginning, causing other disruption expenses. As it was, they formed part of the completion contract without extra expense—other than additional price rises due to the delay.

Bibliography

There is a wide range of literature on the subjects surrounding quantity surveying practice. The list given below is therefore a pointer rather than an exhaustive list. It is unbalanced, partly due to a desire to emphasize some aspects in particular and partly due to a lack of published materials in certain fields. No reference is made to works on quantity surveying techniques other than for cost matters. The articles from professional journals are given especially to help correct this balance. There is an occasional duplication between the articles from journals and the other publications quoted; this may assist those unable to secure one or other of the sources.

Quantity surveying practice

Pre-contract Practice for Architects and Quantity Surveyors. The Aqua Group. (Granada)
Contract Administration for Architects and Quantity Surveyors. The Aqua Group. (Granada)
Tenders and Contracts for Building. The Aqua Group. (Granada)
Which Builder? Tendering and Contract Arrangements. The Aqua Group. (Granada)
    The foregoing books deal particularly with the procedures to be followed for smooth working.
Practice and Procedure for the Quantity Surveyor. A.J. and C.J. Willis. (Granada). This treats especially the matters dealt with in Part 1 of the present work.
Variation and Final Account Procedure. W.H. Wainwright and A.A.B. Wood (Hutchinson Education Ltd). This also gives about half its contents to other aspects and to appendices setting out forms of contract and the like.
Contract Practice for Quantity Surveyors. J.W. Ramus. (Heinemann). This contains a range of numerical examples of detailed work.
Aspects of Civil Engineering Contract Procedure. R.J. Marks *et al*. (Pergamon Press)
UK and US Construction Industries: A Comparison of Design and Construction Procedures. J. Bennett *et al*. (Royal Institution of Chartered Surveyors)
The Anatomy of Quantity Surveying. G.A. Hughes. (Construction Press)
Evolving the Quantity Surveyor. A statement of policy and action. (Royal Institution of Chartered Surveyors)
Future Rôle of the Quantity Surveyor. (Royal Institution of Chartered Surveyors)
Surveying in the Eighties. Policy Review Committee, RICS. (Royal Institution of Chartered Surveyors)
Policy for the Eighties. Annual Conference Report RICS. (Royal Institution of Chartered Surveyors)
A Decade of Quantity Surveying—A Review of the Literature 1970–1979. P.A. Harlow. (Chartered Institute of Building)
Computer Techniques. (Royal Institution of Chartered Surveyors)
Guide to Office Insurance for Quantity Surveyors' Practices. (Royal Institution of Chartered Surveyors)
Study of Quantity Surveying Practice. University of Aston. (Royal Institution of Chartered Surveyors)
Quantity Surveying in the Public Service. (Royal Institution of Chartered Surveyors)
The Quantity Surveyor in the Construction Industry. (Institute of Quantity Surveyors)
Preambles for Bills of Quantities. (Greater London Council)

276

Schedule of Rates for Building Works. Department of the Environment. (HM Stationery Office)

Civil Engineering Bills of Quantities. N.M.L. Barnes and P.A. Thompson. (Construction Industry Research and Information Association)

A Guide to Target Cost and Cost Reimbursable Construction Contracts. J.G. Perry, P.A. Thompson and M. Wright. (Construction Industry Research and Information Association)

Price Adjustment Formulae for Building Contracts. Building Economic Development Committee (HM Stationery Office)

Price Adjustment Formulae for Building Contracts. Quantity Surveyors (Practice and Management) Committee of the RICS. (Royal Institution of Chartered Surveyors)

Formula Method of Price Adjustment. P. Goodacre. (College of Estate Management)

Building Research Station Current Papers. (Building Research Establishment). Several of these deal with operational bills, with performance specifications for building components, and with quantity surveying aspects.

Professional practice generally

The Placing and Management of Contracts for Building and Civil Engineering Works. Report of a committee. (HM Stationery Office.) This is widely known as the Banwell Report, from the name of the chairman of the committee.

Action on the Banwell Report. The report of a working party. (HM Stationery Office). A study of the extent to which the recommendations of the Banwell Report were being implemented during the three years following its publication.

A Code of Procedure for Single Stage Tendering. National Joint Consultative Committee of Architects, Quantity Surveyors and Builders. (RIBA Publications Ltd)

Selective Tendering for Local Authorities. Department of the Environment, R & D Building Management Handbook. (HM Stationery Office)

Early Selection of Contractors and Serial Tendering. Department of the Environment research and development paper on tendering procedure. (HM Stationery Office)

Serial Tendering: A Case Study from Northamptonshire County Council. Department of the Environment research and development bulletin. (HM Stationery Office). Three similar bulletins give case studies of serial tendering in Hampshire, Nottinghamshire and the West Riding.

Tendering Procedures and Contractual Arrangements. E.W. McCanlis (Royal Institution of Chartered Surveyors)

Engineering Services in Buildings: Guide to Tendering Procedure. (Royal Institution of Chartered Surveyors)

Tendering Procedure for Industrialised Building Projects. National Joint Consultative Committee of Architects, Quantity Surveyors and Builders. (RIBA Publications Ltd)

Tender Action—QS Practice Pamphlet No. 1. (Royal Institution of Chartered Surveyors)

Before You Build. National Economic Development Office. (HM Stationery Office)

Develop and Build. Department of the Environment. (HM Stationery Office)

Planning a Major Building Programme. Department of the Environment research and development bulletin. (HM Stationery Office)

The Management of Capital Works Programmes. (National Building Agency)

Civil Engineering Procedure. A report. (Institution of Civil Engineers)

Project Management. Liverpool Polytechnic Department of Surveying. (Institute of Quantity Surveyors)

Nothing but the Truth: Expert Evidence in Principle and Practice. J.A.F. Watson. (Estates Gazette Ltd)

Arbitration for Builders. P.J. Lord-Smith. (Northwood Books)

Accounting for Construction Management. J. Goodlad (Heinemann)

The Insolvency of Building Contractors. Society of Chief Quantity Surveyors in Local Government. (Borough Architect's Department, Brighton)

Architectural practice

The Architect in Practice. A.J. Willis and W. George (Granada)

Handbook of Architectural Practice and Management. Royal Institute of British Architects. (RIBA Publications Ltd)

Working with your Architect. Guidance for clients, by the Royal Institute of British Architects. (RIBA Publications Ltd)

Code of Professional Conduct. Sets out the standard of professional conduct for members of the Royal Institute of British Architects. (RIBA Publications Ltd)

Conditions of Engagement. Royal Institute of British Architects, to regulate relations between architects and clients (RIBA Publications Ltd)

Building law

Building Contracts: A Practical Guide. Dennis F. Turner. (George Godwin) This contains an extensive further bibliography of building and background law, with a list of standard forms.

Building economics and cost control

Building Economics I.H. Seeley. (Macmillian Ltd)

Cost Planning of Buildings. D.J. Ferry and P.S. Brandon (Granada)

Construction Cost Appraisal. C. Dent. (George Godwin)

Building Cost Control Techniques and Economics. P.E. Bathurst and D.A. Butler (Heinemann)

Building Design Evaluation: Costs in Use. P.A. Stone. (E. & F.N. Spon)

Capital Investment Appraisal. C.J. Hawkins and D.W. Pearce (Macmillan Ltd)

Cost Benefit Analysis. D.W. Pearce. (Macmillan Ltd)

An Introduction to Cost Planning. (Royal Institution of Chartered Surveyors)

Building Cost Forecasting. J. Southwell. (Royal Institution of Chartered Surveyors)

Cost Models for Estimating. (Royal Institution of Chartered Surveyors)

Total Building Cost Appraisal. J. Southwell. (Royal Institution of Chartered Surveyors)

Aspects of the Economics of Construction. D.A. Turin *et al.* (George Godwin)

The construction industry in general

Construction Industry Handbook. Edited by R.A. Burgess, P.J. Horrobin, Norman McKee and J.W. Simpson. (Construction Press)

How Flexible is Construction? National Economic Development Office (HM Stationery Office)

Handbook of Information Practice in the Construction Industry. (HM Stationery Office)

Building: the Process and the Product. D. Harper. (Construction Press)

National Working Rules for the Building Industry. (National Joint Council for the Building Industry).
Working Rule Agreement. (Civil Engineering Construction Conciliation Board).

Administration and management in the construction industry

Building Management Notebook. (Chartered Institute of Building)
Construction Projects: Their Financial Policy and Control. Edited by R.A. Burgess (Construction Press)
Contract Planning and Contractual Procedures. B. Cook (Macmillan Ltd)
Modern Construction Management. F. Harris and R. McCaffer (Granada)
Business and Financial Management. B.K.R. Watts (MacDonald and Evans Ltd)
Programming of Building Contracts. Brian Armstrong (Northwood Publications)
Introduction to Building Management. R.E. Calvert (Newnes Butterworth)
Principles of Construction Management. R. Pilcher (McGraw-Hill)
Production and Planning applied to Buildings. R.J. Hollins. (George Godwin)
Management Techniques applied to the Construction Industry. R. Oxley and J. Poskitt. (Granada)
Code of Estimating Practice. (Chartered Institute of Building)
The Practice of Management. P.F. Drucker (Pan Books)
Principles and Practice of Management. E.F.L. Brech (Longman)
Cost Control in the Construction Industry. J. Gobourne (Newnes Butterworth)
Annotated Bibliographies (Chartered Institute of Building)
   There are numerous titles each containing references to areas of material, a number of which relate to sections of this bibliography.
Building Management Abstracts (Chartered Institute of Building)
Building Research Current Papers. (Building Research Establishment) Various titles in this series deal with management and related subjects.

Codes of regulations, scales, forms etc

1 PUBLISHED BY THE ROYAL INSTITUTION OF CHARTERED SURVEYORS
Royal Charter and Bye-laws
Rules of Conduct for Chartered Surveyors
Rules of Conduct: Notes for Chartered Quantity Surveyors
Scales of Professional Charges
Directions to Chartered Quantity Surveyors on Advertisements and Announcements
Standard Form of Quantity Surveyor's Valuation
Building Instalment Certificate.
Definition of Prime Cost of Daywork. This is available in various forms, for building, maintenance, engineering and electrical work, as agreed with the respective bodies concerned.
Form of Agreement for Appointment of Quantity Surveyor.

2 PUBLISHED BY THE INSTITUTE OF QUANTITY SURVEYORS
Memorandum and Articles of Association and Bye-laws.
Scales of Professional Charges.

Articles in 'The Chartered Surveyor', its Quantity Surveying Supplement and 'The Chartered Quantity Surveyor'

(These are classified by topic and date of publication only.)

BROAD POLICY
Public Accountability and the Quantity Surveyor. March 1970
What a Client wants from a Quantity Surveyor. February 1972.
Responsibility and Authority. July 1974
Economic Planning and the Construction Industry. January 1977.
Quantity Surveying: a review of the Challenges. February 1977.
The Changing Nature of Professional Practice. April 1977.
The Construction Industry: its Resources and Market. Spring 1978.

MANAGEMENT AND ORGANISATION
Professional Indemnity Insurance. May 1971
Professional Indemnity Insurance Scheme. December 1976.
The Responsibility of a Chartered Quantity Surveyor in a Multi-disciplinary Practice or Consortium. Autumn 1977.
Quantity Surveying: Practice and Progress. July 1978.

GENERAL PROCEDURES, DESIGN AND ECONOMICS
Construction Contracts: UK v Europe. January 1973
Integration in Design. March 1974
Project Management. Autumn 1974
Insurance and Project Management. Autumn 1974
The QS/Contractor Barrier. Summer 1975
One Statistician's View of Estimating. Summer 1975
'Buying' a New Building: the US Management Approach. Winter 1975/76
To Build or Reconstruct? August/September 1976
Factories: to Refurbish or Redevelop? August/September 1976
The Professional Quantity Surveyor and the Provision of Public Sector Housing. Autumn 1977.
DHSS Expenditure Forecasting Method. Spring 1978.
Operational Planning. Spring 1978.
Financial Feasibility Studies and Design Criteria: the Role of the Quantity Surveyor. Summer 1978.
Cost Models. Summer 1978.
Cost Modelling in Building Design. Summer 1978
Construction Project Control. January 1979
Does Adaptable Design Pay? January 1979
Programming: Responsibilities and Techniques. February 1979
Control Is On The Shop Floor (Aircraft). April 1979
Accent on Value For Money (Petrochemical). May 1979
Overseas Consortia: The Future Role of the Quantity Surveyor. February and March 1980
Ethics in Computer Modelling. May 1981
Planning Costing into Planning. June 1981

Ethical Problems in Computer Modelling. September 1981
CAD—How Beastly for the QS. October 1981
The Nougat or the Toad: Holding Down Production Costs. November 1981
Why Do Tenders Vary? December 1981
Project Cost Prediction. December 1981

TENDERING AND CONTRACTUAL PROCEDURES
The Case for Negotiated Buildings Contracts. March 1971
Firm Price Tenders: a New Approach. December 1973
Formula Variation of Price. Winter 1975/76
The NEDO Formula: Is Total Collapse the Answer? Spring 1977
Nomination: for or against? Spring 1977
Making Good Use of Low Bids. March 1982

PACKAGE AND MANAGEMENT CONTRACTS
The Package Deal and the Professions. September 1973
The Chartered Quantity Surveyor and Package Deal Contracts. December 1973.
Package Deal Services and Fees. October 1981
Management Contracting: Old Relations in New Clothes. October 1981
Management or Fee or Both? May 1982

BILLS OF QUANTITIES AND DOCUMENTATION
The National Building Specification. March 1973
Automated Measurement. July 1975
Towards a Simpler SMM, Summer 1977.
Outline of a Theory of Measurement. October 1979

ENGINEERING SERVICES, CIVIL AND INDUSTRIAL ENGINEERING
Problems associated with Engineering Services Specialist Sub-contracts. May 1972
The Quantity Surveyor in Civil Engineering. March 1974.
The Quantity Surveyor and the Measurement of Engineering Services. June 1976.
Quantity Surveying Involvement in Mechanical and Electrical Services. Autumn 1978.
The Quantity Surveyor in Civil Engineering. January 1979
Cost Control: The Influence of Tendering Methods on Engineering Services Costs. May 1979
Heavy Engineering and the Quantity Surveyor. June 1979
Selecting the Civil Engineering Contractor. July 1979
Engineering Services: Who Prepares the Bill of Quantities? August and September 1979
An Engineer's Reply to the Quantity Surveyor. December 1979
Processing Civil Engineering Claims. November 1980
Quantity Surveying Roles in Industrial Engineering. March 1981
Engineering Services: Quantity Surveying Roles. March 1981
Subcontract Enquiries and ICE5. April 1981
Industrial Engineering: Tendering and Contract Procedure. July 1981
Industrial Engineering: Post-Contract Procedure. July 1981
*The Contractor's Quantity Surveyor in Engineering Services: Management's Need for Clear
Guidance.* September 1981

POST-CONTRACT PROCEDURES
Adjustment of Preliminaries: QS Notes. June 1970
VAT Question Time for Quantity Surveyors. March and April 1973
Post-contract Financial Control. Winter 1974/75
Variations at a Premium. Summer 1977
Post-Contract Cost Control. January 1979
Contractors' Claims. February 1979
Arguing about Rock, Sand and Water. March 1979
Waging War on Contract Delays. March 1980
The Contractor's Use of Bills. July 1980
After the Contract is Over. May 1981
A Systematic Approach to Monitoring Progress. November 1981

LEGAL CONSIDERATIONS
The Law and the Professional Man. May 1971
Towards More Efficient Building Arbitrations. July 1972.
Is Arbitration the Most Satisfactory Method of Settling Disputes? July 1972
A Case in favour of the Standard Form of Building Contract. Autumn 1974
The Surveyor as an Arbitrator and as an Independent Witness. May 1975
The pattern of Current Legislation affecting Buildings. Autumn 1977.
Professional Liabilities. January 1979
JCT, ICE and GC/Wks/1 Compared. April 1979
A Long Way After 'Romalpa'. May 1979
Giving Expert Witness. July 1979

INSOLVENCY OF THE CONTRACTOR
Contractors' Insolvencies and the Quantity Surveyor. June 1971
Bankruptcy: the Legal Aspect. September 1973
Bankruptcy: The Financial Aspect. December 1973
Predicting Failure among Construction Companies. Spring 1978.
Main Contractor Liquidations. June 1979

Articles in 'The Quantity Surveyor'

(These are classified by topic and date of publication only)

BROAD POLICY MANAGEMENT AND ORGANISATION
Project Management? a New Approach. June 1975.
Management in High Risk Areas. September 1976
Graphical Methods for use in Allocating Resources. July 1977.
Contracting: Organisation or Annihiliation. September 1977.
Project Management and the Quantity Surveyor in Australia. November/December 1977.
Avoiding a Dispute. April 1978.
Construction and Quantity Surveying Overseas. November and December 1978
The Quantity Surveyor and His Influence on Europe. January 1979
Project Management. January 1980
The Developing Market for Quantity Surveying Services. July 1980
Management and the Quantity Surveyor. March 1981

A Model for the Design of Project Management Structures. April 1981
The Quantity Surveyor and Project Management. February 1982
Marketing for Quantity Surveyors. March 1982
Does the Client Get What He Wants? May 1982

ECONOMIC CONSIDERATIONS
Self-financing Contracts. November 1974.
What is the Cost of Capital? April 1975.
Financial Decision Making and the Analysis of Risk and Uncertainty. May 1975.
The Time Element. (Two parts; deals with client costs). October and November 1976
Future Development: the Client's Dilemma (Two parts) February and March 1978
Cost Planning in the 1980s? December 1980
Cost Modelling for the Construction Industry. July 1981

TENDERING AND CONTRACTING PROCEDURES AND BASES
Building Contracts with Particular Reference to Negotiated Contracts . January/February 1971.
Cost Plan Tendering using a National Schedule of Rates. March/April 1971.
Alternative Methods of Contractor Selection: Two Stage Tendering. September/October 1972.
New Price Adjustment Formula. May/June 1974.
The NEDO Formula for Adjustment of Fluctuations in respect of Building Works. December 1974.
Formula Method of Price Adjustment of Building Works: Series II. May 1977.
A Suggested Computer System for Package-deal Building Contracts. May 1977.
Negotiated Tenders for Building Contracts in the Republic of South Africa. November 1978
The Right Form for Design and Construct? January 1980
The Cost of Abortive Tendering. July 1980
A Degree of Competition. October 1980
Unconventional Alternatives. February 1981
Design and Build, the Quantity Surveyor's Opportunity. May 1981
Production Oriented Tendering. June 1981

BILLS OF QUANTITIES
A Re-think on Operational Bills. July/August 1973.
The Standard Method of Measurement: Seventh Edition. February and March 1975.
The Application of the SfB System to Building Economics. June 1975.
Preliminaries in Bills of Quantities for Building Works. May 1977.

CIVIL AND INDUSTRIAL ENGINEERING
The Role of the Quantity Surveyor in Civil Engineering Work. March/April 1969
The Role of the Quantity Surveyor in.Civil Engineering. July/August 1970
Civil Engineering Bills of Quantities. July/August 1972.
Method Related Charges and the Quantity Surveyor in Civil Engineering. March 1975
Quantity Surveying and the Petro-chemical Industry. February 1982

POST-CONTRACT PROCEDURES
A Simplified Approach to the Preparation of Interim Valuations and Final Accounts. September/October 1970
A Critical Look at Final Accounts. May/June 1974.

Interim Valuations. (on Sutcliffe v. Thackrah) July/August 1974.
Post-contracts Services in the UK. October 1975.
Plight of the Tertiary Contractor. October 1979
Paying the Nominated Sub-contractor. February 1980
Interpretation of Contract Documents—The Quantity Surveyor's Contribution. April 1980

LEGAL CONSIDERATIONS
Termination of Contract: Its Effect and Consequences. September/October 1970
Insurance in the Construction Industry. July/August 1974.
Insurance and Indemnity Clauses in Contracts. October 1974
The Expert Witness. November 1974
EEC Law. November 1974
Arbitration in the Construction Industry. June 1975
Law of Contract: Offer, Acceptance and Consideration. July 1975
Insuring Inflation during the Reinstatement Period. September 1975.
A Practice Arbitration. (on contract progress) October and November 1975
Performance Bonds in Construction. January 1977.
Contract Insurances. January 1978
Arbitration in Construction and Engineering Contracts. June 1979
The 1979 Arbitration Act. December 1979
Professional Liability Today. June 1980
Ascertaining the Governing Law of a Building Contract. August 1980
Interim Certificates and the Common Law Remedy of Set-off. October 1980
Arbitration and the Courts: Recent Changes. January 1981
The JCT Standard Form of Contract 1980 Edition—Payments. April 1980
Retrospective Time Extensions in Building Contracts. May 1981
Sub-contracting Under the Terms and Conditions of the 1980 JCT Form of Contract. August 1981
Arbitration as Applied to the Construction Industry. May 1982

CONTRACT CLAIMS
Civil Engineering Claims. January/February 1969.
Contractual Claims. March/April 1970 and May/June 1970

INSOLVENCY OF THE CONTRACTOR
Insolvency (Two parts) January and February 1976.
Insolvency in the Construction Industry. September 1977.
Determination of Employment under the Standard Forms of Contract for Construction Works. February 1978.

Articles in other periodicals

Various journals and the like include relevant materials. Amongst these are:
  Building
  The Architect's Journal
  Building Technology and Management
  The QS Weekly
  The Estates Gazette

Index